PHYSICS RESEARCH AND TECHNOLOGY

CRYOGENIC RECEIVERS

Physics Research and Technology

Additional books and e-books in this series can be found on Nova's website under the Series tab.

PHYSICS RESEARCH AND TECHNOLOGY

CRYOGENIC RECEIVERS

SERGIY PUTSELYK

nova science publishers
New York

Library of Congress Cataloging-in-Publication Data

ISBN: 978-1-53619-717-4

Published by Nova Science Publishers, Inc. † New York

Dedicated to: Radiofrequency and RADAR Engineers at FHR

CONTENTS

PREFACE

The proper design of RADARs requires the knowledge of many disciplines including high frequency properties of materials, antenna design, analogue front-end and digital back-ends, signal processing. Considerable knowledge on different types of radar designs was acquired in the last 100 years and in some cases, it is quite challenging to keep an overview in some special fields of radar technologies. One of such fields is so-called *Cryogenic Receivers*, which was intensively investigated over the last 30 years. Many practical solutions were successfully developed and tested. One of the best ways to learn about single components or whole cryogenic receivers front-end and in some cases also part of digital back-end is to study the design choices and resulting performance of previous designs.

This book provides such a review.

In order to cover wide ranges of possible applications, the radars and communication systems with components, placed at cryogenic temperatures, will be considered.

This work was mainly done during author's work at *Fraunhofer Institute for High Frequency and Radar Technique* within hobby group "SQUID/SQIF/Kälteelektronik."

This review book starts with introductory chapter, which is followed by a review on *state-of-the-art* of cryogenic receiver systems. Different types of radar or communication systems as well as the most important cryogenic components will be considered here. Other potential applications like digital SQUID and DROS, mixers on Josephson junction as well as parametric amplifier, which can be used for Quantum Radar, are also shortly mentioned. Though L-Band is mentioned, the cryogenic system could be scaled to other bands, like S- or C- ones.

The author is very thankful to all members of cryogenic team, namely Hicham Barbri, Nadya Ben-Bekhti, Andreas Fröhlich, Oliver Grenz, Felix König, Lukas Naumann, Florian Ralf, Marco Tiesing, Thomas Wirths. Special acknowledgments must be given to Helmut Wilden for his numerous spontaneous ideas, which were very helpful and very encouraging. For the permanent supporting and giving time, the author is thankful to Andreas Brenner. And last but not the least, the author would like to give sincere respect for Dirk Heberling and Peter Knott for their willingness to grasp very complex matters like *Rapid Single Flux Quantum* logics in few minutes during FHR holiday.

Sergiy Putselyk
February, 2021, Wissen, Germany

Abbreviations and Acronyms

ADC	Analogue Digital Converter
ADR	Advanced Digital Receiver (also called All Digital Receiver)
ATT	Attenuator
BPF	Band Pass Filter
CDMA	Code Division Multiple Access
CIR	Circulator
CMOS	Complementary Metal Oxide Semiconductor
DAC	Digital Analogue Converter
DDC	Digital Down Converter
DDF	Digital Decimation Filter
DDS	Direct Digital Synthesizer
DLR	Deutsches Zentrum Für Luft- und Raumfahrt (German Aerospace Center)
DQC	Differential Quantum Cell
DROS	Double Relaxation Oscillation SQUID
DSP	Digital Signal Processing
DUP	Duplexer
ESA	Electrically Small Antenna
EuTEF	European Technology Exposure Facility
FET	Field Effect Transistor
FHR	Fraunhofer Institute for High Frequency Physics and Radar Techniques FHR
FPGA	Field Programmable Gate Array

(Continued)

GESTRA	German Experimental Space Surveillance and Tracking Radar
GSM	Global Special Mobile
HBT	Heterojunction bipolar transistors
HEMT	High Electron Mobility Transistor
HF	High Frequency
high-Tc	High temperature superconductor, Tc is abbreviation for transition temperature between superconducting/ normal conducting state
HPA	High Power Amplifier
HTS	High Temperature Superconductor
HTSSE	High-Temperature Superconductivity Space Experiment
JJ	Josephson Junction
LIM	Limiter
LJJ	Long Josephson Junction
LNA	Low Noise Amplifier
LPF	Low Pass Filter
MEMS	Micro-Electro-Mechanical Systems
MCM	Multi Chip Module
MMIC	Monolithic Microwave Integrated Circuit
MTBF	Mean Time Between Failures
PLL	Phase Locked Loop
PMD	Phase Modulated/Demodulated
PS	Phase Shifter
RF	Radio Frequency
RSFQ	Rapid Single Flux Quantum
RX	general naming for receiving module
sc	superconducting
SFDR	Spurious-Free Dynamic Range

Chapter 1

INTRODUCTION

1.1. GENERAL

Over the last 40 years, one of significant advantages in radar and communication systems was related to application of low temperatures for operation of super- and semiconductor electronic components or antennas. Main improvements have been achieved due to physical effects and technology developments, namely:

- Improvement of *signal-to-noise* ratio [1, 2]
- Application of superconducting (sc) materials in order to reduce the parasitic electrical losses in single elements, e.g., antenna, filters, matching elements between antenna and strip lines, etc. [3-6].
- Development of completely new electronics, which is based on *Rapid Single Flux Quantum* (RSFQ) logics [7-10] or active electrically small antennas based on SQUID/SQIF/bi-SQUID elements [11-14].

Depending on each particular application, one or several single elements were cooled down to low temperatures. This was mainly related to the limitations of cryogenic design, namely availability of cryocoolers with high cooling powers and maintenance free operation at least for one year, space availability for the integration of cryocooler with vacuum vessel inside the RF system. For that reason, applications of cold semi- or superconductor components for the receiving systems were quite limited and were closely related to the development of compact, high power and maintenance free cryocoolers.

In the present work, a review of recent technological developments and practical realization of complete units or single components for radars and communication systems is done. In order to limit discussion, only advances made over the last 20 years are outlined. Due to a wide range of developments, it is possible to find many practical applications. Although the present work does not intend to limit our discussion only to one specific system, it is still very helpful to have some example, which could be realized in the near future. As an "example," an arbitrary system is chosen.

An last but not the least, it is worth to mention that there exists a general unwillingness to use low temperatures, and in order to come to the market, the system must have two features:

- To significantly over perform conventional semiconductor systems: an example is developments done on Advanced Digital Receiver (ADR) systems by HYPRES Inc. Other examples are diverse detector electronic components working on SQUID principles.
- To have comparable size, prices and reliability: classical examples are receivers developed by *Superconductor Technology* and *Toshiba Corporation*.

Therefore, for the successful industrial serial production, the special attention must be given to cryogenic system in order to have it as small as possible, maintenance free operation for at least as one (better three) year, and to have comparable or still acceptable prices. For that reason, in the present work, also special attention is given to the cryogenic systems and some practical designs are considered.

It is worth also to mention that developments of "warm" semiconductor LNA (particular based on HEMT technology, like GaAs & InP) in the last few years have also achieved significant results, for example firms like Caltech, OMMIC, Northrop Grumman und Diramics offer LNA with the noise temperature in the range 5-15 K. For that reason, a very thorough design must be done, before one makes choice between the "warm" and "cryogenic" versions.

1.2. Goals of Present Work

The goals of present book could be summarized as follows:

- to make an overview on the present *state-of-the-art* of applications of superconducting components for radar or communication system,
- to shortly outline possible future developments, for example to propose the S-Band receiver with parallel operations of "conventional" and "superconducting" electronics.

Under the "conventional" subsystem, we imply typical *Cryogenic Receiver*, i.e., one using superconducting filter, semi- or superconducting limiter, cold semiconductor LNA. Under the "superconducting" system, all other versions are implied, e.g.:

- Superconductor antenna, based on microstrip technology,
- Superconductor antenna, based on bi-SQUID/SQIF technology,
- Digital or analog SQUID/DROS antenna and corresponding electronics,
- RSFQ electronics,
- etc.

It is foreseen; the "conventional" and "superconducting" systems are operated independently and have negligible influence on each other.

REFERENCES

[1] Ibarra-Medel E., et al., (2008). 4 Kelvin Cryogenic Characterization of Commericial pHEMT transistors at 9 kHz to 8.5 GHz Range, *J. Low Temerature Physics* 151:418-423.

[2] Bardin J., Ultra-Low-Power Silicon Germanium Cryogenic Low Noise Amplifiers, presentation given at *Cryogenic Engineering Conference*, 13 July 2017, Medison, USA, see also J. Bardin et al. (2017). Silicon Germanium Cryogenic Low Noise Amplifiers, *12th International Workshop on Low Temperature Electronics, IOP Conf. Series: Journal of Physics: Conf. Series* 834:012007.

[3] Shen Z., et al. (1997). Power Handling Capabiltiy Improvement of High-Temperatreu Superconducting Microwave Circuits, *IEEE Trans. Appl. Supercond.* 7:2446.

[4] Lancaster M., et al., (1996). Miniature Superconducting Filter, *IEEE Trans. Microwave Theory and Techniques* 44:1339.

[5] Pond J. M., et al., (1991). Ultra-compact microwave filters using kinetic inductance microstrip, *IEEE Tran. Magnetics*, 27:2696.

[6] Hansen R. C., (2006). *Electrically small superdirective and superconducting antennas*, John Wiley & Sons, Inc., p168.

[7] Mukahnov O. A., (2011). *Hystory of superconductor Analog-to-Digital Converters*, in book *100 Years of Superconductivity*, H. Rogalla and P. Kess, Ed., Taylor & Francis, London.

[8] Gupta D et al., (2012). Modular, Multi-functional digital-RF Receiver System, *Microwave and Optical Technology Letters* 54:847-851.

[9] Brock D., (2001). RSFQ technology: circuits and systems, *J. High Speed Electron Syst.* 11:307-362.

[10] Hayakawa H., et al., (2004). Superconducting digital electronics, *Proc. of IEEE*, 92:1534-1548.

[11] Andrade M., et al., (2015). Detection of far field radio frequency signals by niobium superconducting quantum interference device array, *IEEE Trans. Appl. Supercond.* 25:1603005.

[12] Prokopenko G. V., et al., (2013). Wideband Microwave Low Noise Amplifiers Based on biSQUID SQIFs, *IEEE 14th International Supeconductive Electronics Conference* PE7.

[13] Prokopenko G. V., et al., (2015). SQIF Antenna Measurement in Near Field, *IEEE 15th International Supeconductive Electronics Conference* SQ-P24.

[14] Kornev V. K., et al., (2016). A Guide to Active Antennas Based on Superconducting Quantum Arrays, *IEEE Trans. Appl. Supercond.* 26:1400104.

Chapter 2

REVIEW ON STATE-OF-THE-ART OF RECEIVER SYSTEMS WITH SUPERCONDUCTING ELECTRONIC COMPONENTS

In this chapter, a short review on the preset receiver systems with sc electronics over the last 20 years is done. More attention is given to the systems developed for industrial (serial) applications as well as for ones applied for some specific radars. Some promising detectors or components, which could come into the markets in the near future are also mentioned.

It is worth to note that superconducting electronics developed for satellite and space applications will be only shortly mentioned [1, 2]. Although very interesting and pioneering work was done on many components during several international programs in Europa, India, China, USA, Canada, further progress was limited due to several reasons, i.e., i) missing continuous financial support, ii) very specific application area, which is difficult to apply for *on-earth* systems, iii) limited market.

Table 2-1: List of firms or institutions involved in development of sc components gives a very short summary of some firms or scientific institutions involved in the development of single components or complete systems.

Table 2-1. List of firms or institutions involved in development of sc components

Radar & Communication System (analogue) • Superconducting Technology Inc • HYPRES Inc • Tsinghua University • Tianjin Hi-tech Development Inc. • Cryoelectra GmbH and other institutions • Institute of Physics, Chinese Academy of Science • Toshiba Corporation Inc • University of Birmingham • Chung-Shan Institute of Science, Taiwan	Radar & Communication System (digital) • HYPRES
Components (analogue) • HYPRES: Filters, SQUID/SQIF antenna, • Superconducting Technology: Filter • Tsinghua University: Filter • Tianjin Hi-tech Development Inc: Filter • Institute of Physics, Chinese Academy of Science: Filter • Cryoelectra GmbH and other institutions: Filter • Toshiba Corporation: Filter, phase shifter • University of Birmingham: filter, antenna • SCIRO (Australia): SQUID/SQUIF Antenna, Downconverter, heterodyne oscillator, etc. • And many others ...	Components (digital) • HYPRES: all components of receiver system, components for microprocessor • NEC-ISTEC-AIST (ADC and Microprocessor on RSFQ logic) • NIST Boulder: DAC • MIT Lincoln Laboratory: Memory, analog filters • Raytheon BBN: Memory • IBM: microprocessor architecture • Northrop Grumman: microprocessor architecture • Firms in China and Japan: microprocessors • D-Wave System: microprocessor

2.1. Some National and European Programs

An application of high-Tc superconducting components, e.g., at least for the filters, gives not only advantages of *signal-to-noise* improvement, but also benefits of better filter performance characteristics (e.g., higher steepness), as well as low weights. For example, German Ministry of Educations and Research founded the program “Superconductors and Ceramics for Future Communication Technology” [1] and main goals were:

- demonstration of perspectives of the cryogenic systems, including High Temperature Superconductor (HTS) devices, assembly technologies and cooling,
- evaluation of system benefits as well as comparison between superconducting and normal conducting cryogenic equipments was one important issue; demonstration of the feasibility of cryogenic and HTS technology for commercial applications with emphasis on reliability, reproducibility, and cost effectiveness of the manufacturing processes;
- demonstration of a reliable, efficient, and cost-effective cooling technology.

The program was divided into several sub-projects, e.g.:

- superconductors, novel ceramics, and cryogenics for satellite communication;
- superconductors for the mobile communications
- superconductors and novel ceramics for the directional broadcasting;
- cryocoolers and cryopackaging;
- superconducting materials: manufacturing, characterization, and qualification.

For example, during this program, eight-pole noise reduction filters with bandwidth between 600 to 800 MHz for C-band with insertion losses well below 0.1 dB were demonstrated. A two-stage LNA with operational frequency from 3.2 to 4.3 GHz have noise figure of around 0.5 dB at 77K, compared one with 1.1 dB at room temperature [1].

Another extensive research was done within European Superconducting Systems for Communication (SUCOMS) during 1996-1999, with three key improvement parameters (L-Band operation):

- Increasing of sensitivity and selectivity in a base-station receiver,
- Improving of blocking characteristics in base-station receiver
- Operation of full transceiver with reduced power, size and weight.

Besides high-Tc filters, LNA, mixer, low-noise local oscillators were also tested at low temperatures.

It is also worth to mention that substantial development of cryogenic components for the satellite communications have been also done for other projects, see for example experiment High-Temperature Superconductivity Space Experiment (HTSSE) on ARGOS satellite [2] or ones at International Space Station at the European Technology Exposure Facility (EuTEF) [1].

During the European program "Superconductors and Ceramics for Future Communication Technology" [1], it was found that for the mobile communication systems working in 0.8-2 GHz range, a typical 2 dB improvement is achieved for receiver system working with high-Tc filter and cryogenic LNA in comparison to warm ones. These advantages are related to three improvements:

a) Reduced LNA noise due to cooling (typically 0.8-1 dB at 300K and 0.4 dB at 60K)
b) Reduced physical temperature of the cooled preselection filter (reduction by 0.2 dB for 1-dB insertion loss of filter)
c) Reduced insertion loss of HTS filters in comparison to conventional filters.

It is important to note that for the standard filter specification, contribution *a* and *b* dominate over contribution *c* and therefore noise-figure improvement is mainly due to the lower temperature and not so much due to losses in high temperature superconductors. The situation is changed for the filters with unconventionally steep skirts, where the low losses in HTS are also essential for achieving a low receiver noise figure. For that reason, if low temperature are used for component cooling, the superconducting (sc) filters are practically always applied.

Further very important research results within this program were related to the antenna noise temperature and capacity considerations, i.e.:

- An important condition for transforming a reduced receiver noise figure into the system benefit is low noise temperature of receiving antenna, 150K < $T_{antenna}$ <300K. So, absence of stong noise coming from autos (ignition, etc.), industrial plants, heat sourced with high temperature, is important, otherwise the antenna noise temperature could be very high, up to 2000 K.

- In the noise-limited case, higher sensitivity transforms into an increased coverage, which leads to lower number of base-stations or higher quality of communication channels in strongly shadowed subareas. For areas with higher traffic, the maximum cell radius is not limited by link budget consideration, but rather by capacity consideration (interference-limited system). So, the reduction of receiver noise under certain circumstances can be transformed into increased capacity. This was especially true for CDMA systems, where for a given signal-to-interference-and-noise ratio (SINR), a reduced noise allows a higher multiple-access-noise-level and, therefore, more simultaneous users per area.

And the last but not the least, through many interesting solutions and results were obtained and some prototype systems were built, none of the system came into the marked. Although maintenance free cryocooler for at least of one year was developed, the overall cryogenic system was too "bulky" and expensive.

2.2. Systems with Superconducting Filters and Cold LNA

In this chapter a short overview on the present *state-of-the-art* on receiver systems using the superconducting (sc) electronics as well as antennas based on operation of SQUID/SQIF elements is given.

2.2.1. Superconducting Technology Inc. (STI)

STI company has achieved commercialization of receiving systems, see models "SuperLink-700LTE" [3] and "SuperLink-850G3AB" and in

Table 2-2 below as well as in Figure 2-1. Since 1999, more than 10,000 were sold. Due to applications of sc filters, two benefits are achieved:

- Allows carriers to add 8% or more usable resource blocks and capacity
- Strengthens the receiver's S/N ratio, which can translate into 10-15% or more increased base station coverage area and data throughput.

Table 2-2. Performance comparison of SuperLink700LTE and SuperLink-850G3AB

	SuperLink-700LTE	**SuperLink-850G3AB**
System performance		
Receiver Noise figure	0.8 dB max (from 698.3 MHZ -715 MHz)	0.8 dB, gain – 12 dB
VSWR	1.5:1 Typical	1.5:1 Typical
OIP3	+29 dBm typical	+30 dBm typical
Maintenance	None	None
Reliability	Above 500,000 hours MTBF	Above 500,000 hours MTBF
Filter Response		600-819.5 MHz: Filter 70 dB, Gain 58 dB 819.5-822.5 MHz: Filter 40 dB, Gain 28 dB 849-851 MHz: Filter 30 dB, Gain 18 dB 851-869 MHz: Filter 60 dB, Gain 48 dB 869-894 MHz: Filter 80 dB, Gain 68 dB
Rejection	35 dB minimum at 697.8 MHz	
Rejection	35 dB minimum at 716.2 MHz	
Power Input requirements	48V dc, input power: 50W steady state	80W steady state
Weight	15 lbs	34 lbs
Mounting	According to customer requirements	According to customer requirements

It is worth to note that in order to make compact and low cost version of cryogenic system, several design iterations in order to simplify the cryostat were done, see reference [4] for some older versions of cryostats.

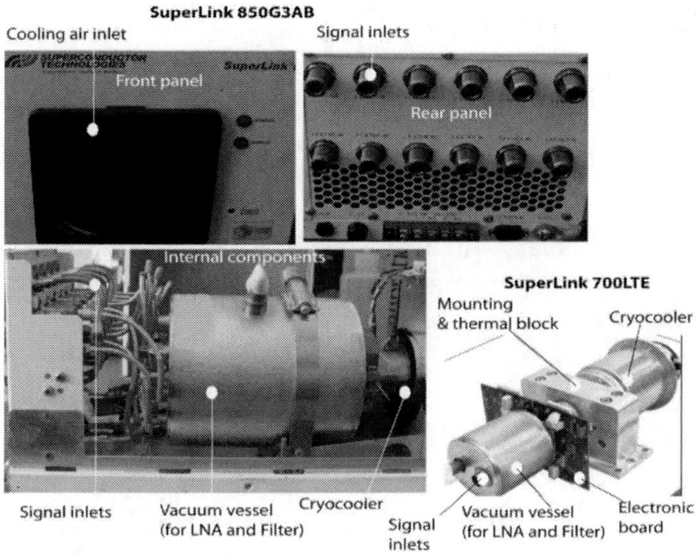

Figure 2-1. Photos of SuperLink 850G3AB (top left and right as well as bottom left) and SuperLink 700LTE (bottom right).

2.2.2. HYPRES Inc.

HYPRES Inc. also developed compact cryostat with LNA and filter, which can be built either from sc or copper materials [5]. It is commercially available.

General Specification:

- Dimensions: 10”·5”·4.”
- Weigth: 2.7 kg
- Frequency: Ka-band (default), X, UHV, Custom
- Input: WR42 square flange
- Output: coaxial K connector
- Operating temperature of LNA: 80K
- Noise Temperature: <50K
- Cool down time: <1 hour
- Peak power: <40W
- Steady-state power: around 25 W
- Recommended maintenance: every 2 years

2.2.3. Tsinghua University

Tsinghua University developed receivers for communication systems. Five such receivers were successfully tested in urban area for 18 months [6]. Due to application of sc filter, it was possible to reduce transmitting power by 2.35 dB.

Filter was 14-poles, and whole receiver was installed in urban area of Tianjin for demonstration and later upgraded to 5 units and installed in Beijing.

2.2.4. Cryoelectra GmbH with other European and Chinese Institutions

Within European program devoted to the development of communication systems for *on-orbit* and *on-earth*, several receivers were constructed, see Figure 2-2 below [1, 7, 8].

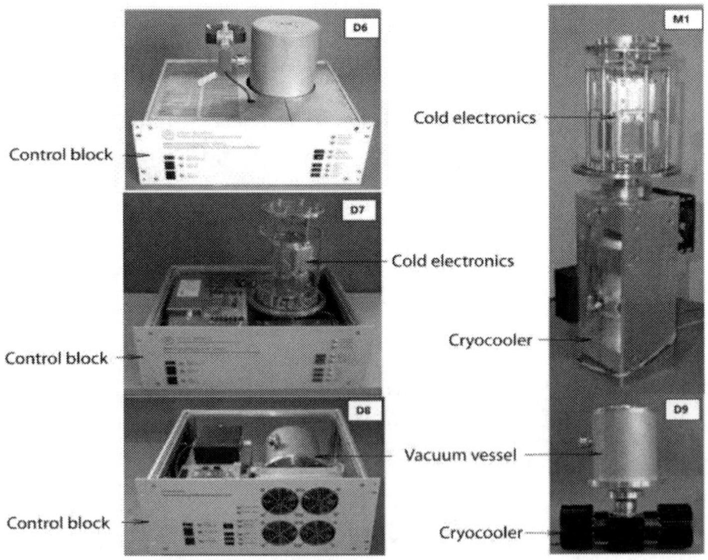

Figure 2-2. Cryogenic Receiver Front End developed by Cryoelectra GmbH in collaboration with other partners: [1, 7, 8].

2.2.5. Tianjin Hi-tech Developments Inc.

Tianjin Hi-tech reported high-performance linear-phase filter with center frequency of 830 MHz with 12-poles and 2 cross couplings for 3G mobile communications. The system was installed in Tianjin in 2004 [6]. Later, this firm also developed tunable 5-pole filter for frequencies 235-250 MHz, in-band losses 2.5-9.4 dB and bandwidth of 0.7 MHz.

2.2.6. Institute of Physics, Chinese Academy of Science

Institute of Physics participated in the development of several radars and communication systems. Figure 2-3 shows the block diagram of transceiver designed for communication in 2009. Due to application of sc material, the system was capable of operation up to 11.7 W and bit error rated was reduced by 80%. Anti-interference capability is improved by 10 dBm and spurious power in emitting signals was restricted by 26 dBm.

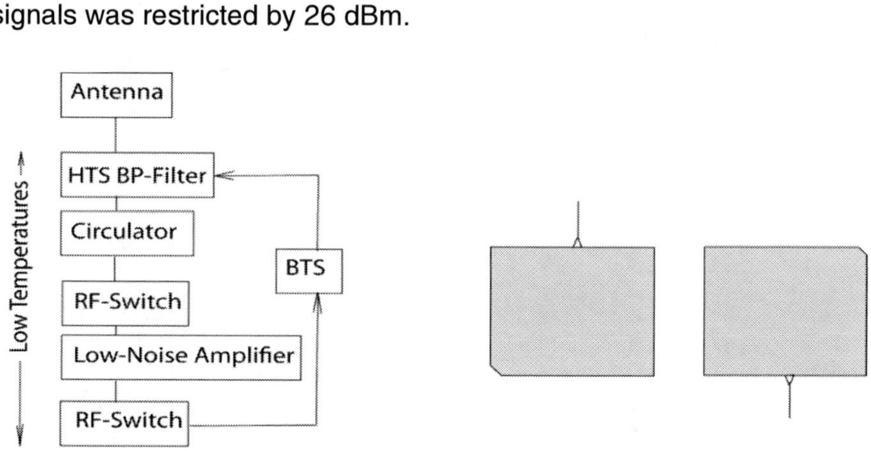

Figure 2-3. Block diagrams of transceiver for communications (left) and a topology of 4-pole dual-mode patch filter (right) [9, 10].

Weather radar was also developed by this institution, see Figure 2-4 below [9, 11].

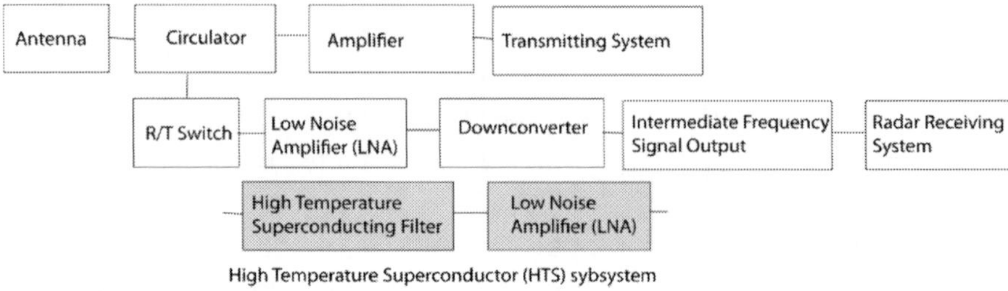

Figure 2-4. Application of sc filter and cold LNA for weather radar: block diagram [287].

Due to application of sc filter the sensitivity was increased by 3.8 dB and interference rejection was increased by 48.4 dB.

For the Miyun Observatory, China, this institute also supplied sc filters for 40 and 50 m diameter telescopes, working in S and X dual frequencies, [6, 9]. The 40 m one was later also used for a deep space station observation.

2.2.7. Toshiba Corporation: Communication System

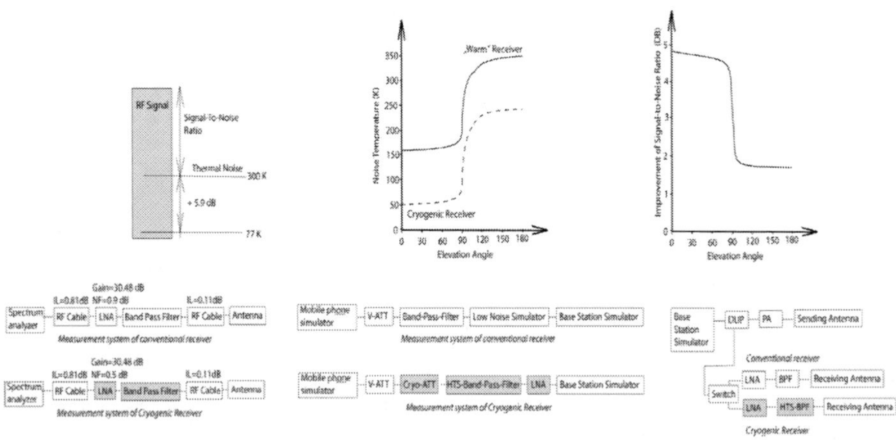

Figure 2-5. Block diagram of measurement system of external environment noise at 2GHz band (left), throughput in cable connections (middle), and throughput by actual radio wave (right) [12].

Toshiba Corporation has developed cryogenic receiver to a mobile phone stations [12], see Figure 2-5.

Table 2-3 shows results of the noisy environment.

Table 2-3. Measured results of RF noisy environment [12]

	Conventional Receiver	**Cryogenic Receiver**
50Ohm termination at 30.8 degrees	-142.80 (dBm/Hz)	-142.80 (dBm/Hz)
Absorver on Horn Antenna	-142.90 (dBm/Hz)	-142.90 (dBm/Hz)
Zenith direction	-146.15 (dBm/Hz)	-147.43 (dBm/Hz)
Horisontal direction	-144.63 (dBm/Hz)	-145.70 (dBm/Hz)

Figure 2-6 shows results of throughputs in cable and in actual receiver.

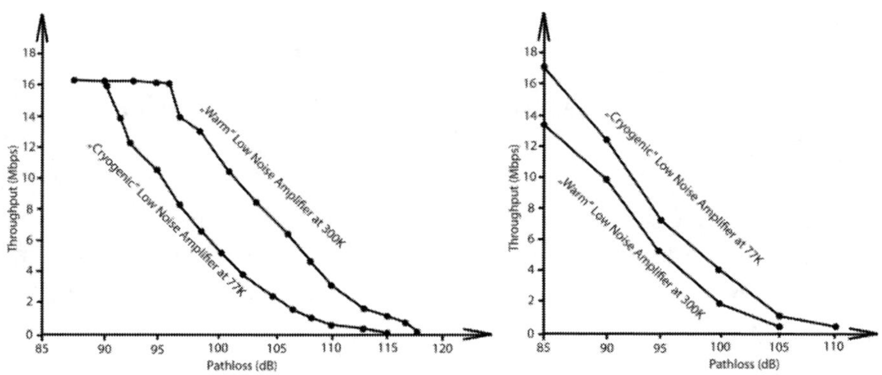

Figure 2-6. Measured throughput in cable (left) and actual by receiver (right) [12].

So, noise improvement was measured to be around 6.1 dB, and throughput one was improved by 1.5 to 3.5 dB [12].

Weather radars for frequencies of 9.7 GHz and 5.3 GHe were also developed by this company, [13-15]. In order to reduce interference, a hybrid filter made from normal and superconducting components was developed. Such hybrid filter was capable of transmitting of 100 kW power with waveguides, while the suppressing side frequencies.

Due to applications of hybrid sc filter the fractional bandwidth was reduced to 0.055% and insertion losses were 1.8 dB. For the sc receiving filter, the bandwidth was 8.8 MHz (0.09%).

Similar results were also obtained for weather radar operating at 5.371 GHz [13-15].

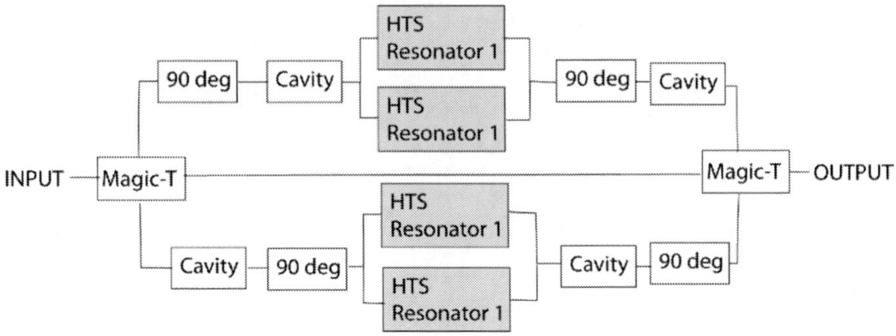

Figure 2-7**.** Weather radar for operation at 5.371 GHz. Transmitting filter block diagram. [13-15].

2.2.8. NTT DoCoMo Inc.

NTT DoCoMo Inc. has developed cryogenic receiver to a mobile phone stations [92], see Figure 2-8. The center frequency was at 1.95 GHz, passband width 20 MHz, sharp selectivity of 20 dB/100 kHz, 1.4 dB ripple, 31.3 dB average passband gain and averaged passband equivalent noise temperature of 47.9 K with total weight of 19 kg and volume of 35 liters.

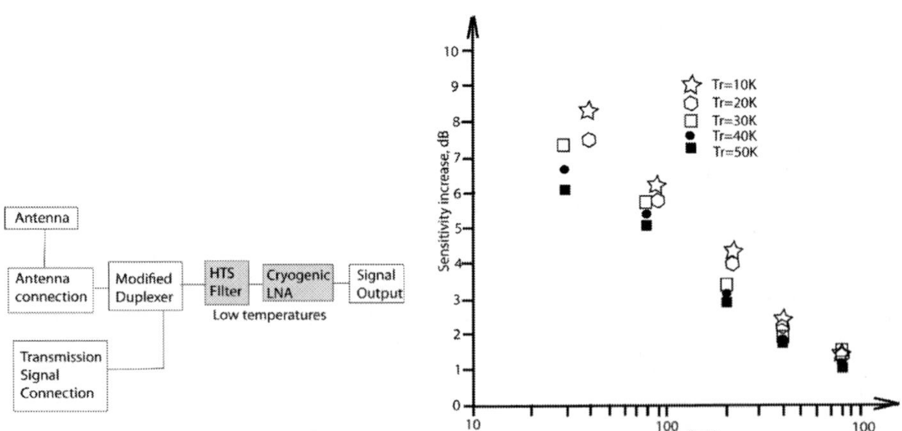

Figure 2-8. NTT DoCoMo Inc. receiver for communication. Left: configuration, top right: sensitivity improvement (*Tr* and *Ts* are cryogenic receiver and antenna temperature, respectively) [92].

Supplementary measurements were done on four identical cryostats with cryocoolers in order to estimate the reliability of the cryogenic systems. For four cryostats, total operation time of 170,000 h without failure has been achieved, which implies that 98 from 100 cryostats will achieve Mean Time To Failure of 40,000 h or five years.

2.2.9. University of Birmingham

University of Birmingham participated in the development of sc antennas and filters. Many sc filters were supplied for testing or for some applications, e.g., in radio astronomy, e.g., Jodrell Bank Radio Observatory and Multi-Element Radio Linked Interferometers Network working in the frequency range 400-MHz till 2 GHz [16-20].

2.2.10. Chung-Shan Institute of Science, Taiwan

Chung-Shan Institute of Science, Taiwan also developed a sc filter for tracking and searching radars, see Figure 2-9 below [21].

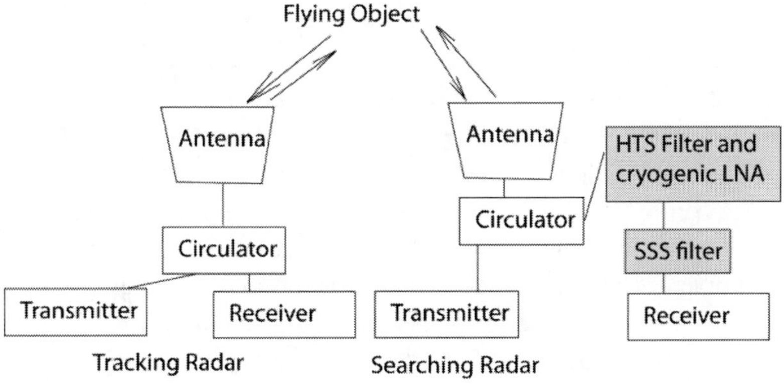

Figure 2-9. HTS filter for tracking and searching radar: schematic of radar [21].

2.2.11. Radars and Communication Systems for Space Applications

Worldwide several programs have been devoted to the development of cryogenics components for the operation at satellites, e.g.:

- USA: High Temperature Superconductivity Space Experiment (HTSSE-I and HTSSE-II phases), with testing on-orbit in 2000. Main result is that cryogenic components are applicable for satellite communication and surveillance systems for military and civil applications. Component testes were (only some of them are mentioned): channelizers/Filters, receivers, A/D converters, digital IFM, Digital Multiplexer, delay line, antenna array [2].
- France: developments of some components within project “Superconducting Cryogenic Active Receiver Element System” [22].
- India: Developments of some components [23].
- Canada: developments were aimed to: i) A 60-channel HTS multiplexer similar to requirements of INTELSAT 8 program, ii) A Ka-band “Beam Link” subsystem that is similar to requirements of multimedia satellites. Program was financed by canadian DARPA, TRP, CSA, DND [24].
- Germany: Program “Superconductors and Ceramics for Future Communication Technology” (1998-2002) and one of five subprojects was devoted to systems for the satellite communications. Main result is that cryogenic components are applicable for satellite communication and surveillance systems for military and civil applications. Practically all components were developed and tested on the earth [1].
- China: sc filter development and on-orbit testing [6, 9, 11, 25].
- Israel: cooling of different components at satellites [26].

2.3. ARRAY ANTENNA WITH SC FILTERS AND OTHER COLD ELEMENTS

Toshiba Corporation (Japan) made significant progress in development very compact transceiver systems with phased array antenna. In additional to the sc filter, several other components working at cryogenic temperatures were developed. The antenna was not cooled and was designed for quick dis/assembling possibility.

Though application of superconducting materials for antenna will surely give some more gain or lower parasitic losses, copper (or silver) plated micro-strip antennas give some practical advantages, like:

- No shifting of resonance frequency due to shrinkage of materials at low temperatures. At low temperatures, plastic materials shrink by 2-5%, which leads to reduction of element sizes and inter-element spaces (metals, like copper, stainless steel or silver shrink by 0.2-0.3%). Surely, it is possible to design cryogenic micro-strip antennas with limited shrinkage, though it is a challenge due to space limitation inside the vacuum vessel.
- Easiness of antenna change: it is not necessary to open the vacuum vessel just to change antenna, for example for the tests of other geometries.
- Better utilization of available space in front of antenna for receiving/transmitting of signal. In case of cold antenna, vacuum vessel somehow occupy available free space between each module, so weaker signals have to be measured or directivity of antenna is changed.

2.3.1. S-Band 8 Element Transceiver Module with Cold Bandpass Filter and LNA

One of the first phased array receiving systems, operating at 3 GHz, which used high-Tc band-pass filter and cryogenic LNA, see Figure 2-10 till Figure 2-11 below, was developed by Toshiba

Corporation [27, 28]. Receiver unit had 8 channels with high-Tc superconducting filter (inter-digital hairpin microstrip line resonators with sizes 22×29.5 mm), which was developed at Toshiba Corporation, as well as cold LNA. The receiver unit had dimensions 152×180×119 mm. Insertion losses of high-Tc filter were below 0.1 dB, and noise temperature of this receiving unit was less than one third as compared to one of only LNA at room temperature, i.e., around 100K at surrounding temperature of 77K and 330K at ambient temperature of 295K.

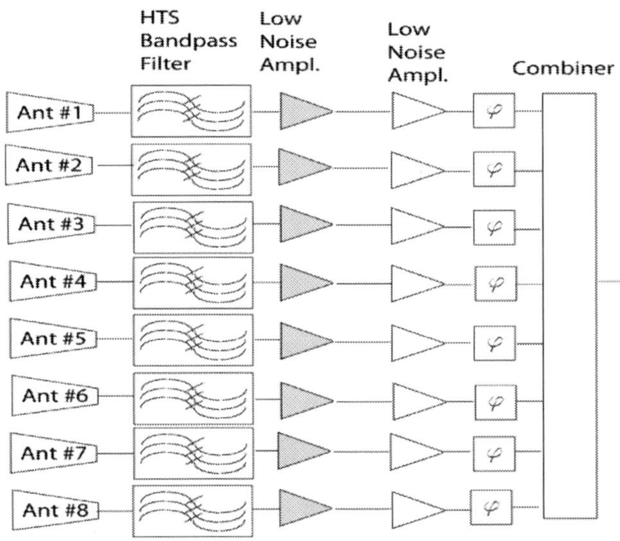

Figure 2-10. Block diagram [27, 28].

The transmission response and main losses of cryogenic unit at 77K were around 20 dB and -15 dB (max), while reflection losses of HTS filter were below -20 dB.

The main parameters of most critical components could be summarized as follows:

- High-Tc superconducting Filter: 5-poles Chebyshev, YBaCuO thin film on MgO substrate, manufactured by photolithography and dry etching. It consisted of inter-digital hairpin microstrip line resonator, and the resonator was half-wavelength microstrip line with inter-digital capacitor at the both end. 3dB bandwidth was around 1 GHz, insertion loss in the ripple bandwidth was smaller than 0.10 dB.

- LNA: 2-stage MMIC LNA was manufactured from GaAs FET. The bias condition was drain voltage of 6 V and gate voltage of 1.10 and 1.35 V. This condition is minimal Noise Figure at 295K. Gain is improved by 2dB by cooling to 77K and noise figure was also improved by 1.8 dB.
- Modules: the amplitude and phase differences between single modules were 0.45 dB and less than 1°, respectively. Noise Figure difference between single cryomodules was less than 0.1 dB.
- Cylindrical Active Phased Array Antenna: 16 elements of half-wavelength dipole antenna was realized. Antenna pattern was measured in azimuth between -90 and 90°. Beam scan angle was decided by phase conditions of each 6 bit phase shifter (PS) in the combiner part
- Gain for frequency range 1000-2000 MHz was 20 and 22 dB for operation at room and cryogenic (77K) temperatures.

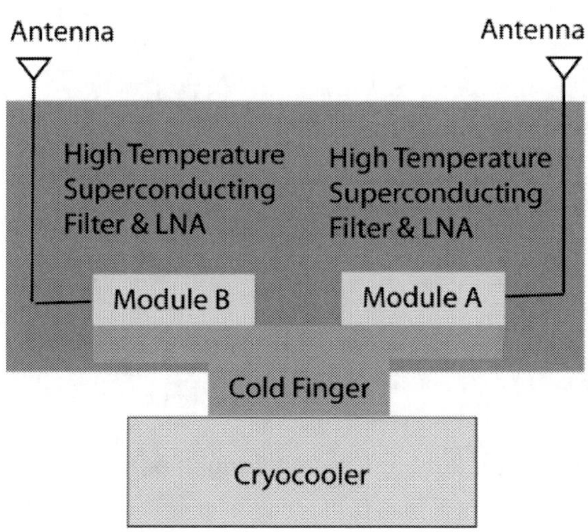

Figure 2-11. Block diagram of receiving unit with 8 modules [27, 28].

Cryogenics: small single stage cryocooler with 10W cooling power at 77K. For reduction of heat load through RF cables, 10 cm copper-plated SUS coaxial cables was used. Estimated losses were 0.3 W at 295K.

2.3.2. 16 Element Transceiver Module with Cold Limiter, Bandpass Filter and LNA

Further developments of S-Band antenna were developed to three main areas [29, 30]:

- Combining of receiving/sending (transceiver) in one single module, in order to use such single module (also called "sub-array module") for larger arrays.
- Increasing number of elements to 16 per single sub-array module
- Application of microstrip antenna, e.g., patch one, instead of cylindrical dipole antennas.

Figure 2-12 show block diagrams of transceiver and receiver parts of multichannel sub-array antenna.

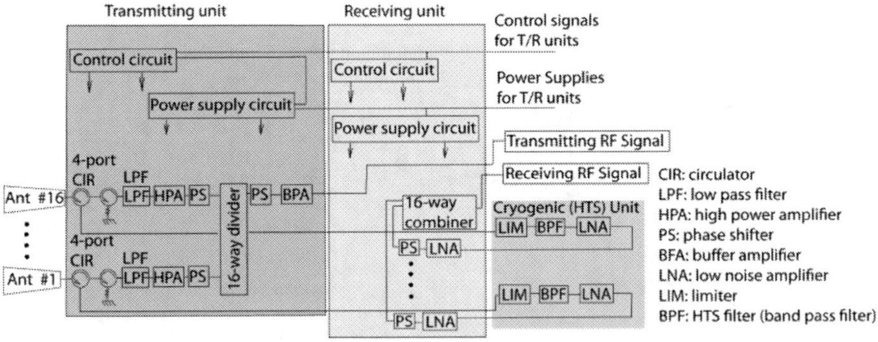

Figure 2-12. Transceiver (left) and receiver (right) parts of multichannel sub-array antenna [29, 30].

Very remarkable approach of assembling in "sandwich" style of single components was applied.

Two receiving systems were very similar and main difference was in design of RF interface at the vacuum vessel (it is worth to note that typically this "interface" is done with SMA/SMP/G3PO connectors). In one case, this was "bulky" version, and in other case was "flat." So, it was possible to reduce the height of vacuum vessel. It was also possible to have slightly different design of LNA.

The outside dimensions of receiver unit were 200×240×435 mm, and electronically beam scanning range: azimuth – 90°, and elevation – 60°. Design of high-Tc filter and LNA was very similar to the previous ones, see chapter 2.1 for more details.

One of the most critical components was receiving circuit outside the vacuum vessel. It included 16 phase shifter, a control circuit, a power supply circuit and a combiner circuit of RF signals.

Figure 2-13 shows schematic view of components of receiver unit.

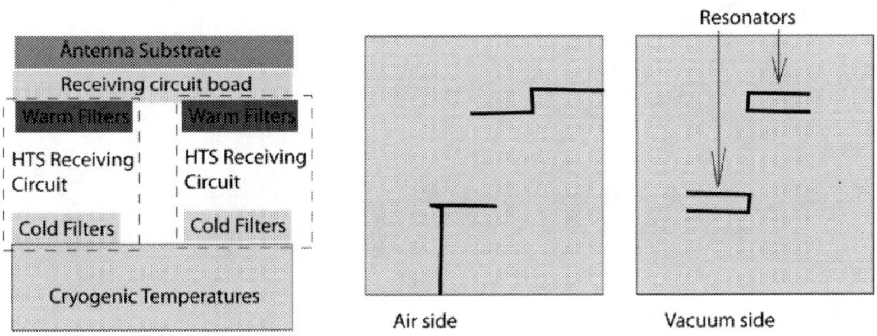

Figure 2-13. Schematic view of receiving unit (left) and receiving circuit at warm side (right) [29, 30].

2.3.3. X-Band 16 Element Transceiver Module with Cold Bandpass Filters and LNA

Another example of a similar system, developed by the same Japanese corporations, is shown in Figure 2-14 below [31]. The transceiver design was very similar to the previous one with main difference in operation frequency, i.e., 9 GHz with 1 GHz bandwidth. Electrically beam scanning range was: azimuth – 90°, elevation – 60°.

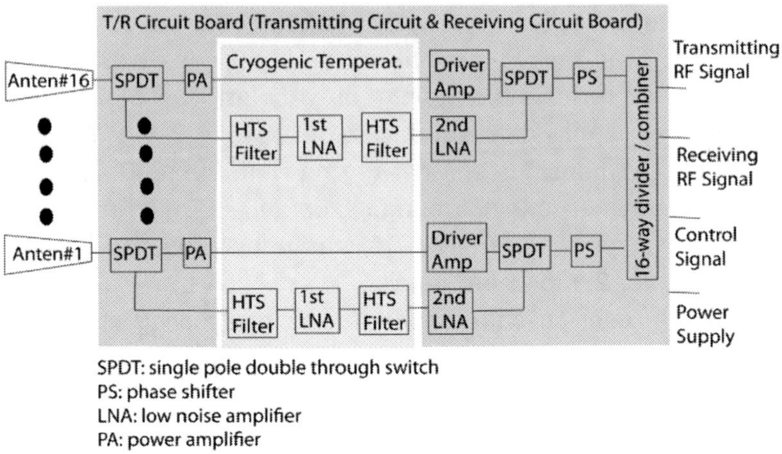

Figure 2-14. Block diagram [31].

The configurational views of transceiver are shown in Figure 2-15.

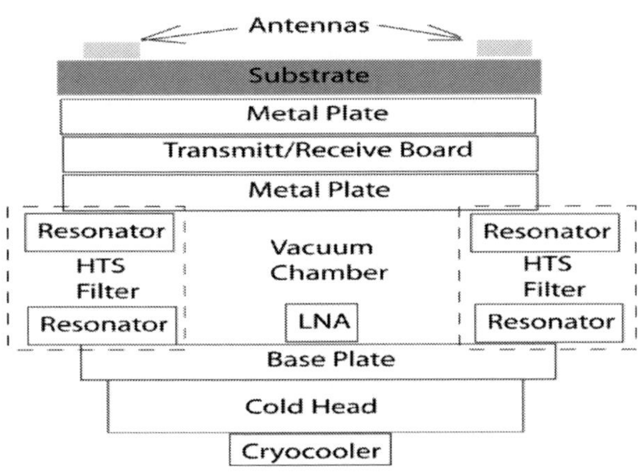

Figure 2-15. Configuration views of transceiver unit [31].

For the effective coupling of RF signal from antenna to filter at receiving unit, a filter consisting of two parts with gap of 1mm in between was designed. Upper part consisted of copper and lower of high-Tc filters.

In the frequency range 8.7-9.5 GHz, the Noise Temperature was in the range 50-90K for cryogenic receiver and 90-130 for warm one [31].

The cooling capacity increased from 2 till around 2.15 W for duty cycles increase from 4 till 16%.

Saturated output power for 16 ports was in the range 26-29 dBm [31].

2.3.4. X-Band 16 Element Transceiver Module with Cold Phase Shifters, 16 Way Divider/Combiner, Tx-Rx Switch, Limiter and LNA

A complete Rx&Tx module with 16 element array for X-Band (9 GHz), which was developed by Toshiba corporation [32], could serve as a good example of scalable phased array antenna, see Figure 2-16 below.

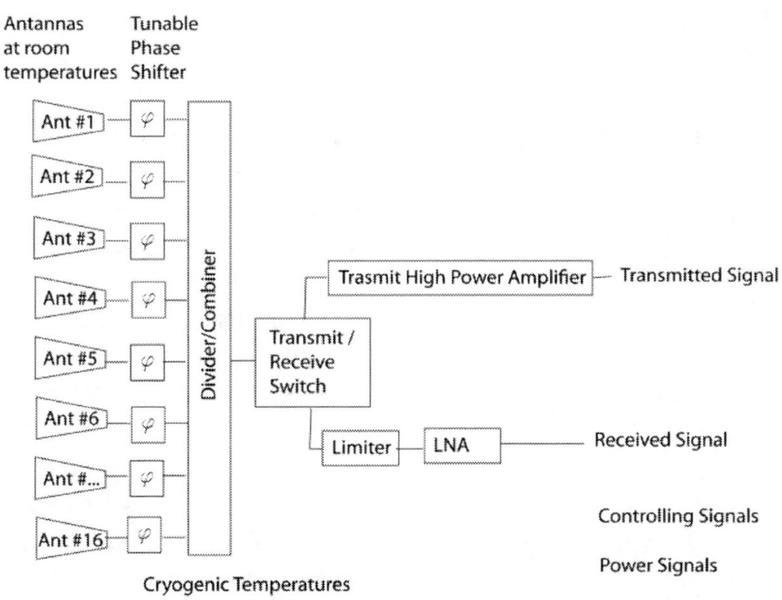

Figure 2-16. Schematic view of phased array antenna and single module [32].

The number of warm antenna elements was 16 (4×4) and dimensions of sub-array module – 80×80×330 mm. Electronic beam scanning range was: azimuth – 60°, elevation – 60° and noise temperature, which included antenna noise temperature of 89K (at elevation angle of 2°), was 315K.

Figure 2-17 shows the schematic view of phase shifter and Tx-Rx switch.

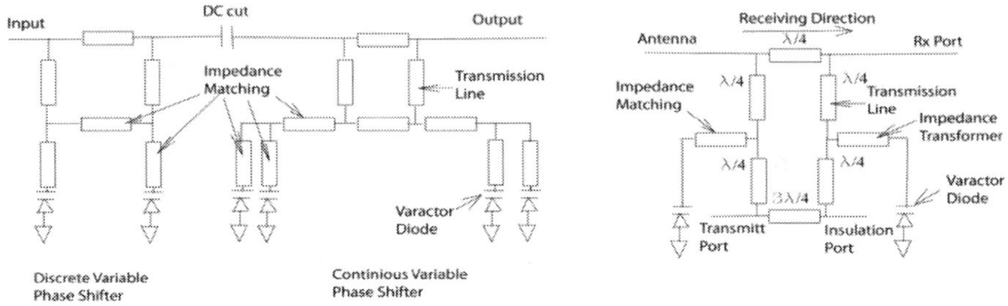

Figure 2-17. Schematic view of phase shifter (left) and of Tx-Rx switch (right) [32].

Parameters of the most critical components could be summarized as follows:

- Phase Shifter: it consists of discrete variable phase shifter, which makes 0-180° phase shift and continuous one that makes fine adjustment, see Figure 2-17 (left). All transmission lines make from high-Tc YBCO on MgO substrate, and have very low insertion losses (less than -0.9 dB). Varactor diodes are used for varying phase shift and overall power consumption of one phase shift module was around 11 μW. Phase control voltage from -18 V to 0V and mode change (0° /180°) allows variation of 360° phase shift.
- Tx/Rx switch, see Figure 2-17 (right): all transmission lines are made from high-Tc YBCO on MgO substrate, and have very low insertion losses (less than -0.3 dB in Rx mode (S21) and -0.5 dB in Tx mode (S31)). Tx-Rx insulation is around 25 dB and the leakage signal from Tx port to Rx port can be sufficiently suppressed by a limiter to protect LNA. In Rx mode, negative voltage is applied to both varactors, while in Tx mode, no voltage is applied.
- 16-way divider/combiner: superconducting technique was also applied but since a resistance cannot be created on a MgO

substrate, a thin film chip resistor was connected for an isolation resistance by bonding wires. The insertion loss were less than 0.05 dB for 2-way divider/combiner.

Gain was in the range of 12-18 dB and noise temperature 250-700K.

2.3.5. Superconducting Antenna, Filter and CMOS-Based LNA

A compact front-end receiver for frequencies 2 and 10 GHz was developed group of Prof. K. Yoshida, see Figure 2-18 [33, 34]. Such system has several details, which are important for the future development of superconductor analog electronics, i.e.: i) compactness, ii) low costs, e.g., CMOS-based LNA, iii) sc filter, antenna, matching network.

Figure 2-18. Schematic diagram of superconducting front-end receiver [33, 34].

Such compact and potentially low-cost receiver systems have potential for wide applications.

2.4. Chapters 2.2 and 2.3: Lessons Learnt, Future Trends and Research Activities

Lessons Learnt:

- Cryogenic system worked reliable over many years. This expectation is somehow against “feeling” of many engineers working with radars and communication systems. Operational experience on large series, like STI Inc., or single prototype exemplars showed that cryogenic system was usually operable for more than 4 years (for many cases with typical MTBF for STI cryocooler of 800,000 till 1M hours) without maintenance [4, 36, 37, 92]. It was quite typical, that receiver/transceiver with cryogenic components was replaced due to brakeage of other components, e.g., small air fans or electronic components like power supply units, or modified because of upgrading the conventional semiconductor electronic components.

- In order to come to the market with serial product, the cryogenic system must be as simple as possible. Nice and very informative example is SuperFilter/SuperLink series of STI Inc., where low cost and reliable cryocooler was developed as well as several massive activities were undertaken for cryostat simplifications.
- For serial filter production, miniaturization is not required, because i) total volume is defined by cryocooler and cryogenic LNA and ii) tuning and trimming is more difficult for low dimension filters.
- It is possible to design any sc filter for each particular applications, e.g., very narrow band, band-stop, band-pass, notch, multi-band, any combination of band-stop/band-pass/notch filters.
- SC filter and cold LNA shows significantly better results than semiconductor electronic components for two cases: i) if antenna temperature is below 50K, i.e., for astronomy, searching and tracking radars, satellite communications from ground stations, and ii) large interference in very closed frequency range, i.e., due to excellent skew characteristics, such interference is efficiently reduced. The second argument is practically the only one for application of STI receivers for the terrestrial communication systems [35].
- For terrestrial communication systems, it is still difficult to judge whether the applications of sc filter with cold LNA is advantageous in comparison to conventional technologies. This is mainly related to a lacking of clear comparison between these two competing systems as well as detailed investigation of operation of the sc components, namely:

- For the sc filter and cold LNA system, it would be helpful to distinguish the contribution of sc filter, e.g., reduction of interference, reduction of insertion losses, and LNA (e.g., larger linear and dynamic ranges).
- To measure sc electronics in especially dedicated system, i.e., without "insertion" in the conventional system (also called as "intruders"). In this case, better performance of sc receiver is expected.
- For some communication systems, e.g., GSM, the improvement is marginal, while for others, e.g., CMDA, UMTS could be significant. This is related to the different operational principles, namely the GSM system is much less sensitive to the absolute level of sending/receiving signal, while for CMDA it is crucial to have very sensitive detection.
- Operation strongly depends on the interference signals, e.g.: i) possibility to suppress unwanted signal, ii) communication net topology, e.g., interfering signals from neighbor or other nets.
- In urban areas, the antenna temperature is in the range of 150-250K, but in some cases could be up to 2000K, e.g., in the near of energy sources, autos, factories, fires, etc., So, two systems have very similar performance, because overall system temperature is dominated by the antenna noise temperature.
- Geometry of surface areas, e.g., hills, large buildings, are also not properly described or not know, so it is difficult to judge experimental results related to the quality of connections.

It is worth to note that for the terrestrial communication systems, it is somehow believed that in average few dB could be gained due to applications of superconducting filter and cold LNA. However, due to lacking of all above mentioned boundary conditions, for the customer it is difficult to calculate whether applications of cryocooled receiver is beneficial in comparison to the conventional one.

An the last but not the least, the sc filter working with powers of 10-100 W were developed, which make them suitable for the transceiver system for the terrestrial communication systems.

2.4.1. Future Work, Trends and Research Activities

For terrestrial communication systems, it would be very helpful to have detailed comparison of communication systems based on conventional and cryogenics components for different operational scenarios, e.g., urban/rare region, large/small interference, variable land surfaces, different communication protocols, etc.

As the most important trends and research activities, it is worth to mention three topics:

- Development of sc filter with higher critical temperatures, e.g., based on Thallium thin films,
- Tunable Filters are very promising for special applications, e.g., for filter with very narrow (0.5% or below) bandwidth and large tunable range, e.g., 10-30%.
- Active impedance matching between an antenna and other components as well as active tuning of superconducting antenna, e.g., by aperture tuning by switch. Over the last 10 years, this field was actively developed for warm antenna (particularly with the small geometries) in order to have i) smaller losses in matching network, and ii) to tune antenna to different frequency ranges. Similar work can be done for transceivers with cryogenic components.

Though the development of even higher temperature sc tunable filters are ongoing, the active impedance matching and antenna aperture tuning were not extensively investigated. It is worth to specially mention two groups in Japan, namely, one leaded by Prof. K. Yoshida at Kyushu University for the development of low costs and compact receiving system with sc antenna, matching network, filter and CMOS amplifier, see chapter 2.4.5 for more details, as well as several groups at Toshiba Corporation for the development of many superconductor based components, see chapter 2.3.4. It could be possible that with the development of diverse superconducting components or complete systems with sc antenna, filter, matching network, low temperature semiconductor LNA; all benefits of superconducting analogue electronics will be considered for practical systems and more applications will be realized.

It is also worth to keep in mind that micro cryocooler family (mainly Stirling type) are permanently under development, and some versions, e.g., K561 or K562S of Ricor company, are very compact (40-75 cm^3) and light (145-290 g). Though MTTF of these cryocoolers is still relative low (10,000-12,000 hr), it is possible to apply them for development of mobile receivers working on sc technology.

2.5. HYPRES Inc. Systems with SC Electronics Based on RSFQ Logic

In this chapter a short overview on the present *state-of-the-art* on receiver systems using the superconducting digital electronics is given. Such electronics is based on operation of Rapid Single Flux Quantum (RSFQ) logics. Due to limited available space, no details on the operation of single elements will be given, and interested reader should contact papers and books written by three authors and their groups, who made a significant contribution to this field, namely Oleg Mukhanov, Vasilii Semenov and Konstantin Likharev [38-45]. It has to be also mentioned that significant and quite impressive work on RSFQ logic, e.g., microprocessor, ADC, etc., was done by several japaneese groups, e.g., Hisao Hayakawa at Nagoya University, and others at Yokohama University, International Superconductivity Technology Center (ISTEC), etc. [46].

And the last but not the least, it is worth to underline that although many papers, books and review articles on description of Josephson Junction (JJ) operation or basic operation of RSFQ logic have been published, a complete and detailed description on *state-of-the-art* design of RSFQ circuits is still missing. Situation is more complicated for devices and systems, where only few papers, which were often published over several years ago, have been presented in the opening literature sources. This is mainly related to the two facts, i.e., i) only limited number of systems is produces for very specialized applications ("user-customized"), and, therefore, not all information is available in the opening literature sources, ii) several systems have been developed for military applications. Nevertheless, for engineers, it would be very helpful to have some books describing RSFQ components and devices, as well as literature sources with more detailed description of the present *state-of-the-art* radio frequency systems.

2.5.1. General Description

The *All Digital Radio* (ADR) systems, developed by HYPRES Inc, use design based on "digital-RF" approach, which means that signal conversion and signal processing is done at RF rather than conventional baseband [47-63]. Input wide-bandwidth RF signal is directly applied to ADC modulator, which is producing oversampled low-bit-width code at speed in the range of 20 to 50 Gbps. Before down-conversion, this stream is further processed in low-complex but high speed and throughput processor (DSP). Following functions are done (if needed), e.g.: a) digital signal combining from different channels, b) true-time delays for digital beamforming, c) adaptive active cancelation of transmit channels in case of application of RF-Transmitter, d) correlation-based digital filtering, e) spectrum management and control, etc.

Thereafter, the signal is down-converted to baseband by using digital mixers and decimation filters for further processing and decoding. The analog-to-digital (ADC) converter and DSP can operate at similar clock rate up to 50 GHz. Synchronization with external low-rate clock (MHz - few GHz range) is typically done with Phase Locked Loop (PLL), which is operated as smaller frequencies than clock one. If it is necessary, before ADC, an analog filter or low noise amplifier (LNA) could be implemented (according to the company information, analogue filter before ADC is practically always applied, though it is seldom shown in diagrams). Digital decimation filters are applied to decrease the output bandwidth, while increasing the effective number of bits. Signal can be divided into main and quadrature components or be channelized into multiple baseband channels.

For the transmitting, the similar operations are performed, see Figure 2-19 and Figure 2-20 below. Digital-to-analog (DAC) converts signal to RF band, which is further supplied to high power module for final amplification. A dynamic digital equalizer, which is typically integrated within DAC, is used for compensation of nonlinearities in high power amplifiers. Digital addition of different channels with different carries frequencies leads to a one broadband signal and as consequence to only one high power amplifier. Additional advantage of high frequency DAC/ADC and processing unit is that RF predistorter can correct instantaneous, signal-dependent fluctuations of the HPA transfer function in picosecond time range. So, superconducting electronics will perform RF predistortion in the digital domain [60].

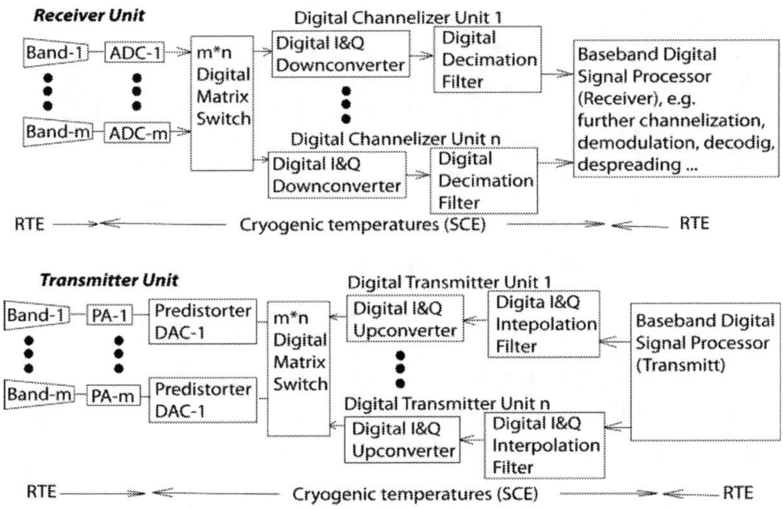

Figure 2-19. Block diagram of Digial-RF transceiver with components located at room temperature (RTE) and cryogenic one (SCE). For the simplicity of presentation and understanding, several antenna are shown, though only one is installed [53].

High power amplification, back-end processing and antenna are placed at room temperature. Low Noise Amplifier (if necessary) could be placed either at room or at cryogenic temperatures, while components for analogue filtering (if necessary) as well as duplexing are typically placed at cold (4K) or intermediate one (40-60K).

In the following chapter, previously built and tested *All Digital Receiver* (ADR, and sometimes called as *Advanced Digital Receiver*) systems are shortly summarized and some most critical components are discussed in more details. It is worth to mention, that due to limited information in the opening literature sources, some rough calculations will be done in order to estimate the system or component performance.

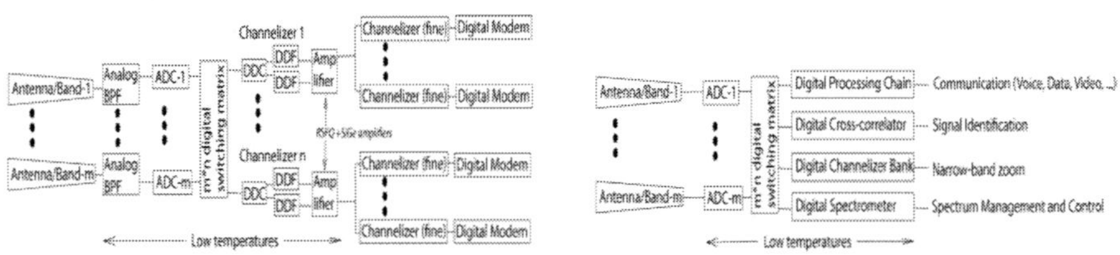

Figure 2-20. Multi-band, multi-channel digital-RF channelizing receiver architecture. Each digital (coarse) channelizing unit (or more general – digital processors) comprises a digital downconverter (DDC) and two digital decimation filters (DDF). For the simplicity of presentation and understanding, several antenna are shown, though only one is installed [56].

2.5.2. First Generation

In 2004, the first Digital-RF channelizing receiver chip was produced at HYPRES. It had of a 20 GS/s L-band first-order low-pass phase modulated/demodulated (PMD) ADC modulator operated up to 24.3 GHz clock rate. Data transfer to "warm" electronics (FPGA) was at 95 MSps. In 2005, this chip was integrated into cryostat with small cryocooler. This word's first Digital-RF receiver system was named as "ADR-0" [54, 64].

The first practical system (ADR-001), which was a complete X-band satellite receiver prototype, was delivered to the Joint SATCOM Engineering Center (JSEC) in Ft. Monmouth, NJ and demonstrated direct digitization, digital down conversion and filtering of X-band (7.25-7.75 GHz) signals from XTAR, DSCS and the newly launched WGS satellites [49, 55, 56, 64]. The chip had X-band second-order bandpass Σ-Δ ADC. It is worth to note that the RF signal could be also digitized and without down-conversion at low temperatures was transferred to room temperature electronics [57]. The receiver was successfully integrated with a digital I&Q MODEM from L-3 communications, demonstrating demodulation of a MILSTD- 188-165A uncoded BPSK-modulated waveform at 1.544 Mbps [ca. 40 MSps] with maximal rate at 6.5 Mbps, see Figure 2-21 below. In 2007, the ADR-1 system was upgraded with installation of 30 GHz chip [55].

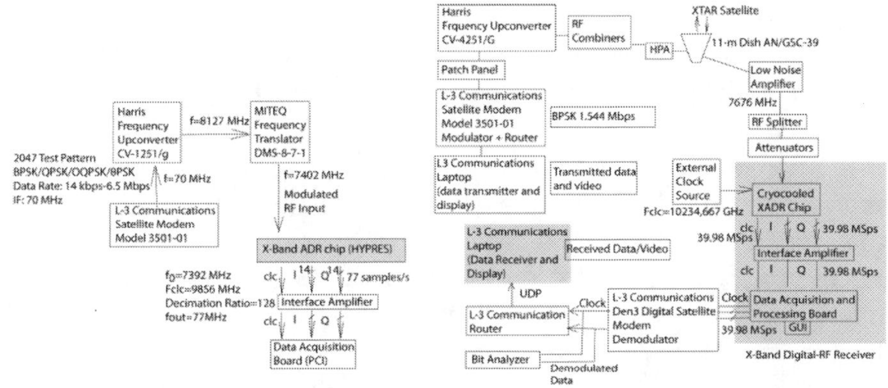

Figure 2-21. Test configuration for the X-Band ADR-001 chip with a carrier frequency of 7.402 GHz [56] (left) and life test with XTAR satellite (right) [53].

A very similar second system (ADR-002) was delivered to the Air Force Research Laboratory. It hosted a low-pass digital channelizing receiver chip, clocked at 24.32 GHz, targeting SIGINT functions for bands in the lower RF range [64].

Due to modular designs, the chips of these systems were later upgraded for the operation at higher frequencies.

2.5.3. Second Generation

Two second generation systems, i.e., ADR-03 and ADR-04, used upgraded chips, optimal placing of components inside the cryostat, e.g., high-temperature superconductor filter was placed at 70K temperature level, modification of thermal radiation shield, etc. Further significant progress was achieved in the data transfer rate between cryogenic and room temperature components. Operation of 600 Mbps for 9.6 GHz clock frequency allowed using smaller deserialization ratio smaller and less cables. Systems typically used low-pass PMD ADC as well as band-pass Σ-Δ ADC [49, 64].

In January 2009, teams from HYPRES, ViaSat, and Navy demonstrated the world's first multi-net Link-16 data link in which analog outputs from two *Link-16* terminals, operating with independent frequency hopping patterns, were combined as well as encrypted *Link-16* analog waveforms for two MIDS (Multifunctional Information Distribution System) terminals and supplied to L-band band-pass Σ-Δ ADC chip integrated with RSFQ 1:16 demux, see Figure 2-22 below [49]. Two modified MIDS terminals, each sending its hop-net sequence to the ADR-004 and receiving the digitally dehopped data, were used for the demonstration. Each receive terminal could be programmed to receive either of the two transmitted Link 16 networks. Digitized RF data were deserialized by a factor of 16 on the superconductor chip and by another factor of 4 on the FPGA before performing two-step channelization. The first step of channelization divided the band of interest into four 75 MHz wide sub-bands (SB-1 to SB-4), each represented as 16-bit complex (I and Q) words at 150 MHz. For two-net reception, two copies of these coarse channelized data are produced for further processing through two independent, identical paths, comprising a switch and a fine channelizer. The switch, upon receiving the hop sequence, selects one of the four sub-bands containing the current hopped carrier frequency for its output. Next, the fine channelizer translates the carrier to baseband by performing complex multiplication between the I and Q words and the corresponding frequency offset data that is generated and stored for all the carrier frequencies. An analog HTS filter was added on the first (40 K) stage to reject unwanted parts of the spectrum. Although these chips work at a clock frequency of at least 25.6 GHz, in ADR-004 the clock frequency was limited to about 10 GHz due to the cross-talk between the data outputs at the vacuum feedthrough. Starting with the specified baseband clock rate of 5 MHz, the ADC sampling clock frequency of 9.6 GHz was chosen to obtain an integer decimation ratio (30) together with a power-of-two deserialization ratio (64).

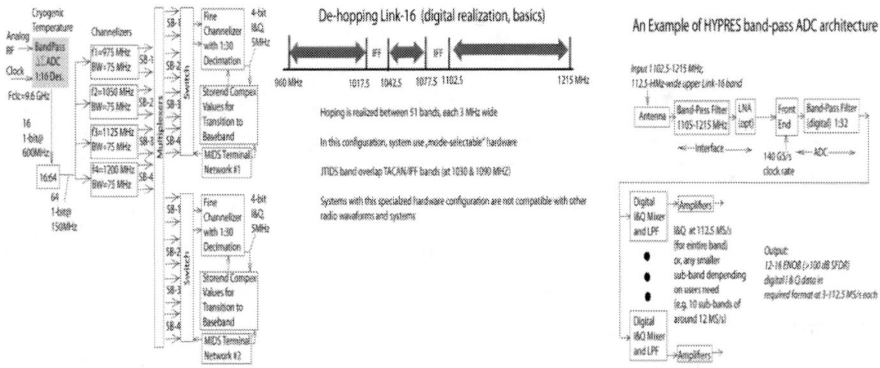

Figure 2-22. Processing scheme with coarse and fine channelization for Link 16 [49, 65].

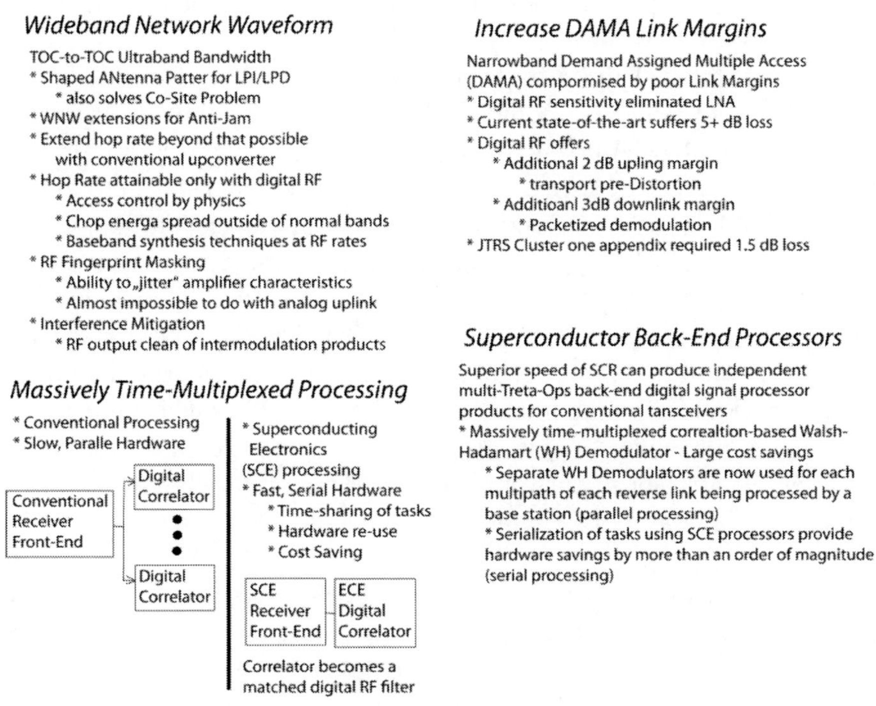

Figure 2-23. An example of application of ADR system for commercial communication systems [65].

This system also had some advantages for the commercial communication systems; see Figure 2-23 above.

2.5.4. Third Generation

This generation made step towards fully digital receivers [49, 64]. Third-generation ADR systems was designed to house two chip modules with 80 high-speed digital I/Os, and up to four 17-channel interface amplifier assemblies. The design was modular and allowed independent servicing of each chip module and quick field replacement if necessary. Due to modular design, the system reconfigurability enabled ADR-5 to be configured and used for satellite communication tasks. Operation with the XTAR satellite without utilizing a low-noise amplifier, which is usually required for such kind of applications, was demonstrated. ADR-5 system worked up to 32 GHz clock frequency with a variety of phase and amplitude modulated waveforms at 30 Msymbol/s data transfer rate.

In 2009, the ADR-005 system, which was initially designed for X-band SATCOM use, was completed. One of the goals was to repeat the operation of ADR-001 system, done in 2007. The satellite signal frequency (f_{RF} = 7.676 GHz), waveform (uncoded BPSK-modulated data at 1.544 Mbps), ADC sampling frequency (clock frequency $f_{clk} = 4f_{RF}/3 = 10.234667$ GHz), and decimation ratio (256) were chosen to be identical. In contrast to the 1 cm^2 X-band channelizing receiver chip demonstrated on ADR-001 system, a 5×5 mm^2 X-band band-pass Σ-Δ ADC chip was used in conjunction with a polyphase channelizer implemented on a commercial Xilinx Virtex-4 FPGA. The interface to the L-3 communications digital I&Q MODEM with 16-bit I&Q words at $f_d = f_{clk}/256$ = 40 Mbps was unchanged. Figure 2-24 below shows a simplified measurement scheme also equipped with conventional analog-RF receiver.

ADR-005 was also operated without LNA, i.e., signal from satellite was directly applied to ADC. Without LNA, no significant performance degradation was observed since the noise pedestal from the satellite's power amplifier was still above the quantization noise floor of the ADC. This chip was also tested at higher clock frequencies (24-32 GHz) with diverse phase and amplitude modulated waveforms at 30 Msymbol/s, see Figure 2-25 below. The total deserialization factor was 32, including a factor of 2 in the room-temperature data interface module.

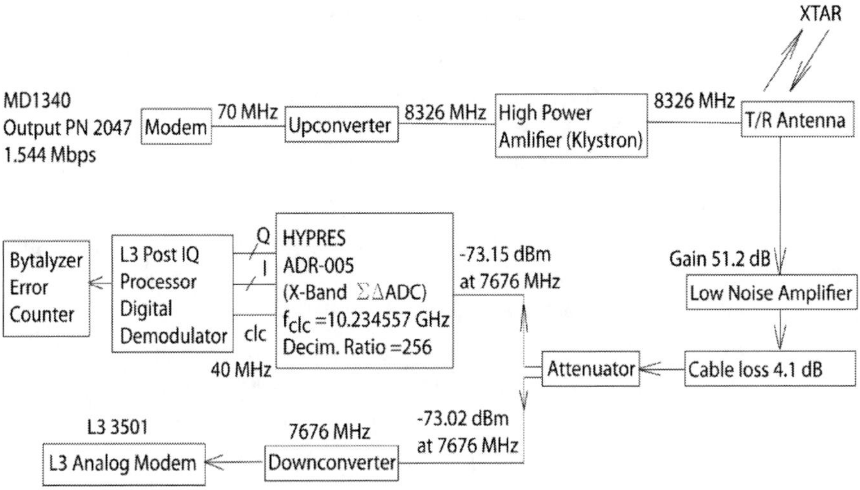

Figure 2-24. ADR-005 operation in SATCOM mode (schematic diagram). Conventional analog-RF receiver is also shown [49].

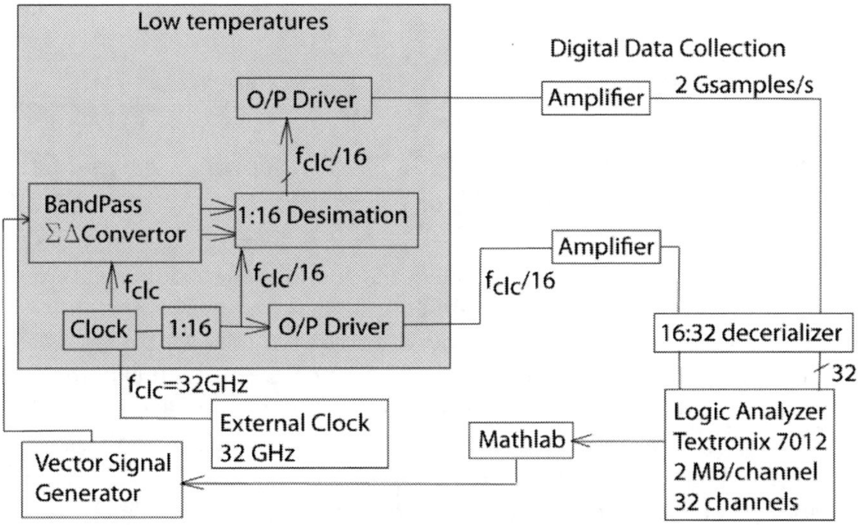

Figure 2-25. ADR-005 under tests at different clock frequencies [49].

It is worth to note that at that time the digital channelizing receiver approach has been extended to include multiple ADC modulators and multiple channelizer units on a multi-chip module [47, 49]. It is also possible to integrate a multi-band but single-channel digital-RF receiver on a single (presently standard) 10 x 10 mm^2 chip by integration of multiple ADC modulators to a single channelizer. A chip integrating four ADC modulators (centered at 850 MHz, 4 GHz, 7.5 GHz and 12 GHz), 1x4 digital switch matrix and 1:16 demultiplexer was demonstrated in 2010 [47, 49]. Due to digital filtering, it was possible to operate ADC either with limited band width (for narrow signals) or with wide band width (in order to scan all bands). It is also worth to note, that it is possible to make an epoxy glued double (or more) chips (so called multi-chip module (MCM)) on single 10×10 mm substrate [47]. Operation of diverse ADC on substrate with different critical currents was also demonstrated. Another example is a multi-chip containing 2×2 switching matrix for two (band-pass at 850 MHZ and low-pass for HF/VHF) ADC and two channelizers with 1:16 deserializers [47]. In the broadcast mode, digitized data from any ADC was copied to both channelizers, while for cross (bar) mode, the outlets from ADC were routed to corresponding channels. Similar chip was developed for X (7.5 GHz) and Ka (20 GHz) bands [47].

2.5.5. Fourth Generation

Official web-site of HYPRESs Inc (https://www.hypres.com/products/advanced-digital-rf-receiver/).

> "HYPRES is proud to announce its fourth-generation Digital-RF Receiver product. Using state-of-the-art superconductor analog-to-digital converter (ADC) technology, the ADR can directly digitize RF (no analog down-conversion). This modular cryocooled, 19-inch rack-mounted system is completely free of liquid cryogen and is capable of digitizing multi-GHz bandwidth with a center frequency up to 21 GHz.
>
> The HYDR-02 is completely customizable and upgradable. In its basic configuration, HYDR-02 includes one user selected superconductor chip. HYDR-02 can be upgraded and reconfigured by purchasing additional chip modules. Additional cryogenic electronics, such as low-noise amplifier, analog filter, may be incorporated within HYDR-02."

Table 2-4 below summarizes technical parameters.

Table 2-4. General information on HYDR-02 system

Features	Applications	General Specification	Installation & Warranty
• Directly digitize RF (no analog down-conversion) • Ultra-wide bandwidth • Multi-band, multi-carrier operation • Compatible with any digital processor • Turnkey operation • Modular upgrades with new chips as they become available	• Software Radio • Spectrum Monitoring • Satellite Communication • Wireless Communication • Electronic Warfare • Radar • Digital Dehopping • Channelizing Receiver	• Size: 22" wide, 30" deep, 78" high • Power: 110V, 60 Hz, single-phase, 2 kW • Weight: 400 lbs • Operating Temperature: +5C° to 35°C	• On-site installation and training • 1-year standard warranty • Recommended maintenance: Every 2 years

The most important *configuration options* are mentioned below.

1. Chip Module (it is possible to select one or two with different clock frequency and I/O configuration)
    a. Single-ADC BPΔΣ (fRF: 0.8 – 21 GHz)
    b. Single-ADC LPΔ
    c. Dual/Tri/Quad ADC (up to 8 synchronous ADCs)
    d. Custom
2. Crycooler
    a. Sumitomo SRDK-101D, 0.1 W@4.2K
    b. Sumitomo SRDK-1.5 W@4.2K
3. Interface Amplifiers
    a. 17-Channel PECL/LVDS Amplifier unit (28+ Gbps per unit)
    b. 17-Channel BlueBird Amplifier unit (90+ Gbps per unit)
    c. Custom
4. Current Source (Model CS-48-100) (Select up to two current sources, must match chip requirements)
5. FPGA Output Interface Board
    a. Xilinx Virtex 6/7/Ultrascale (plus)
    b. Custom
6. Accessories/Add-ons

a. Cryoelectronics Module (T = 40-80 K, temperature controlled)
b. Rapid Deflux Apparatus
c. Low-profile (48" high, 22" wide)
d. External clock source (20 GHz, 40 GHz, 50 GHz)
e. Expert system for automated setup control and remote operation
f. PC with touch screen monitor with configured GUI
g. Vacuum Pump
h. Digital Data Recorder (Record 14+ GHz instantaneous bandwidth for hours)

Though relatively little information is given in the open literature sources, we tried to summarize technical parameters below.

Other technical parameters, which are found in the HYPRES Inc. web-site https://www.hypres.com/foundry/niobium-process/, see also HYPRES Brochure "World-Class Foundry Services," are summarized below.

a) High-Resolution Analog-to-Digital Converter (Phase Modulation-Demodulation) Oversampling modulator clocked at 28 GHz and programmable digital decimation filter: 89 dB SNR at 10 MHz BW, 100 dB SFDR
b) Bandpass Delta-Sigma Analog-to-Digital Converter (various bands, including L, C, X, Ka bands): Up to 32 GHz Clock, 50-65 dB SFDR
c) Flash Analog-to-Digital Converter: 20 dB SNR at 20 GHz BW
d) Ultra-sensitive Analog-to-Digital Converter: 1 microA full-scale range with SQUID front-end
e) Memory RAM: 32-bit, 16 Kb, 100 ps cycle time
f) Shift Register: 19 GHz, 1 Kb
g) Digital Correlator: 40 Gbps 1-bit autocorrelator
h) Time-to-Digital Converter: 33 ps (30 GHz) multi-hit time resolution
i) On-chip Clock Source: 10-1000 GHz SFQ clock, 5-10 fs time jitter
j) Multi-chip Modules: >50 Gbps interchip data transfer

2.5.6. Most Critical Components

Table 2-5 summarizes operational frequency of components working on RSFQ logics for 2-3 μm Nb technology for the year 2004 [44].

Table 2-5. Demonstrated operational frequencies of RSFQ components for 2-3 μm technology (state-of-the-art for 2004 year) [44]

Components	Operational frequency
Toggle flip flop	144 GHz
4-bit shift register	66 GHz
6-bit Flash ADC	3 ENOB at 20 GHz
14-bit High-resolution ADC (2 MHz)	14 ENOB & -100 dBc SFDR
1:8 Demultiplexor (synchronous)	20 Gb/s
1-bit Half-adder	23 GHz
8*N bit serial multiplier (parallel-to serial converter)	16 GHz
128-bit autocorrelator	16 GHz
2-bit counter	120 GHz
1-kbit shift register	19 GHz
6-bit Transient digitizer with 6*32 bit on-chip memory buffer	16 GS/s
19-bit Digital-to-Analog converter	Fully functional at low speed
1:2 Demultiplexor (asynchronous)	95 Gb/s
2-bit Full-adder	13 GHz
14-bit digital comb filter	20 GHz
Time-to-digital converter	31 GHz

2.5.6.1. ADC

Practically all superconducting ADC fall in two categories, i.e.: Nyquist-sampling ADC as well as oversampling ADS, and typically subdivided in three classes: Flash Converters, Counting Converters, which are unique due to superconducting technology and belongs to a class of Delta & Sigma-Delta converters, and finally, classical Delta & Sigma-Delta Converters.

In Nyquist-sampling ADC, the sampling frequency is twice the Nyquist one, which is present in signal (ideal case), or somehow a bit higher in order to consider also performance characteristics of filters. Most commonly used Nyquist-sampling ADC is the Flash Converter, where a large number of separate quantizers (single-bit comparators) each representing single quantization level are built. In many cases, the performance of this ADC type is limited by the precision of quantization level, which is limited by resistor values in the resistor network (semiconducting ADC). The comparator bank feeds a logic circuit that generates a coding for each voltage range. Direct conversion is fast but with limited number of bits (just for example, 8 bit converter needs 256 comparators). In case of superconductor Flash ADC, the design complexity is strongly reduced due to usage of the SQUID comparator. Periodical transfer function in units of single flux leads to the fact that for n-bit superconductor flash ADC only n clocked SQUIDs are required. Like semiconductor ADC, the superconducting ADC is susceptible to many design parameters like mismatched circuit components, resistor ladder, delays in clock and signal paths, local magnetic environment, and thermal fluctuations.

Counting converters utilize the ac Josephson effect, i.e., voltage-to frequency conversion. The quantization and sampling are done separately. First, a Josephson junction produces a single flux quantum pulse sequence, with repetition rate proportional to the applied voltage (quantization step). For that, typically small resistor at the inlet converts the current to the voltage, and inductance between this resistor and Josephson junction prevents propagation of single flux quanta in the opposite direction. So, small resistor, inductance and Josephson junction forms a SQUID loop, which also acts as an integration step as well as implicit (i.e., "naturally built-up") feedback. Next, the signal (sequence of single flux quanta) is counted over some interval. This method of analog-to-digital conversion is similar to low-pass first-order sigma-delta (ΣΔ) modulation. The fast toggle flip-flops, which are ripple converters, count signals. Finally, an averaged digital voltage over counting time is saved. In addition, several other examples were successfully realized, e.g., i) classical dc-SQUID, which is powered to the voltage stage, and input coil for the signal, or ii) direct injection of signal into dc-SQUID (flux quantizing or tracking ADC design).

In the oversampling ADC, the signal is sampled at a frequency, which is much higher that the Nyquist one (typically by using single quantizer). The digital filtering and feedback techniques are applied to decreased the quantization noise and enhance the effective dynamic range. These ADC are typically divided on delta (Δ, differentiation step) or sigma-delta (Σ-Δ, integration and differentiation steps) converter.

In the delta-encoded ADC, the signal has up-down counter, which supplies signal to digital-analog converter. The input signal and signal from digital-analog converter are compared in comparator. The comparator controls the counter, and whole circuit has negative feedback from comparator to adjust the counter till signals from digital-analog converter and original one are equal. So, signal is read from counter and could be integrated (if necessary). In the Σ-Δ converter, the oversampled signal is filtered to the required signal band. Thereafter, the signal is converted to analog one and is subtracted from the original one. Due to negative feed-back, the noise shaping error occurs (due to flash ADC) and final digital filtering for filtering of undesired signal as well as for the improving of the output resolution is performed. The Σ-Δ ADC is the best architecture for direct band-pass sampling, because it shapes the noise around a center frequency, which leads to maximal SNR, while Δ ADC are the best suited for low-pass filter.

Typically, counting ADC design is based on a phase-modulation-demodulation technique, which is unique for rapid single flux quantum (RSFQ) technology. The derivative of the analog signal modulates individual delays of single flux quantum (SFQ) clock pulses within each clock period. The phase demodulation is performed by a race arbiter producing 2- to 4-bit code depending of the number of arbitration units.

The second-order Σ-Δ ADC with centering frequencies around 4 GHz, 5 GHz, 7.5 GHz and 20 GHz were developed, while for 1 GHz, only first order Σ-Δ ADC was designed in 2008.

It is also possible to apply RF undersampling method with some reduction on the sensitivity. The clock frequency must be slightly higher than the center frequency, by factor 4/3, and digital local oscillator with frequency difference between clock and center one, must be applied for the down-conversion.

2.5.6.1.1. Flash-Type ADC

Table 2-6 summarizes recent developments in Flash-type ADC and Figure 2-26 shows the summary results of different types of Flash and multi-bit comparators.

Table 2-6. FLASH ADC

Year & Ref.	Group	Parameters	Comments
2011 [68]	Japan (International Superconductivity Technology Center, Tokio; AIS; Tokyo Denki University)	Operation at 30 and 50 GHz sampling rate, 4-bit effective resolution.	Due to integrated error correction and interleaving circuit, operation with 5-bit was possible
2013 [69]	HYPRES	Operation at 20 GHz sampling rate, 4-bit effective resolution	Three different types were used
2013 [70]	HYPRES	Operation at 30 GHz sampling rate, 8-bit hardware and 4-bit effective resolution.	Three different types were used
now	HYPRES	20 dB SNR at 20 GHz BW	

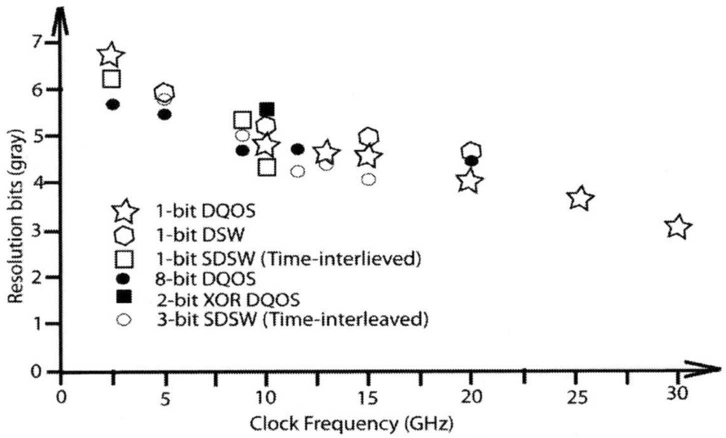

Figure 2-26. Summary of different types of flash comparators and different multi-bit ADC [70].

2.5.6.1.2. Phase Modulation-Demodulation (PMD) Oversampling ADC

Figure 2-27 shows the functional model and Table 2-7 summarizes resent results of PMD ADC.

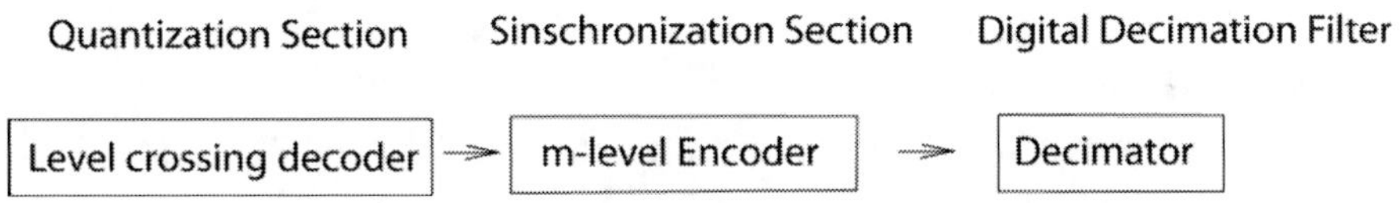

Figure 2-27. Functional model of ADC based on phase-modulation-demodulation scheme.

Table 2-7. PMD ADC

Year & Ref.	Group	Parameters	Comments
2007 [71]	HYPRES		A quarter-rate PMD ADC was tested at clock rate of 81.92 GHz.
2009 [72]	HYPRES	89 dB SNR at 10 MHz BW, 100 dB SFDR.	Clock is at 28 GS/s and during multi-rate operation 46 GS/s. Effective 14.5 bits.
now	HYPRES	89 dB SNR at 10 MHz BW, 100 dB SFDR	Clock is at 28 GHz and programmable digital decimation filter

2.5.6.1.3. Bandpass Δ-Σ ADC

Table 2-8 summarizes resent results of bandpass Δ-Σ ADC. It is worth to note that similar to semiconductor ADC, the superconducting ADC are using time-interleaving, subranging, cascading methods in order to improve overall performance.

Table 2-8. Delta-Sigma ADC

Year & Ref.	Group	Parameters	Comments
2007-2010 [47, 49, 50, 53, 55]	HYPRES	SNR was in the range of 28-60 dB and bandwidth 880-500 MHz (ENOB: 4.5-6.4) depending on decimation ratio	Ka-band. Clock is in the range of 30-32 GHz, Signal is at 7.6 GHz
now	HYPRES	89 dB SNR at 10 MHz BW, 100 dB SFDR	Various bands, L, C., X, Ka: up to 32 GHz clock, 50-65 dB SFDR

2.5.6.1.4. Time-to-Digital Converters (TDC)

TDC is counter based and operated by counting of SFQ clock pulses between two successive input events (hits), so each hit reads accumulated counts out, resets the counter, and shifts the data (or time stamps) to multi-hit buffer.

Hypres Inc Company had also developed Time-to-Digital Converters, see Table 2-9 below.

Table 2-9. Time-to-digital converters*

Year & Ref.	Group	Parameters	Comments
2002 [73]	HYPERS	100 ps time resolution	With an additional prescaler, it is possible to obtain time resolution to 5 ps [74]
2003 [75]	HYPRES	6-ps time resolution and 30-ps multi-hit time resolution	
now	HYPRES	33 ps (30 GHz) multi-hit time resolution	

* The *time resolution* is the precision of time interval measurement or LSB (least significant bit), while the *multi-hit time resolution* is the capability of a TDC channel to measure a few time intervals sequentially.

A combination of Time-to-Digital and Analog-to-Digital converters was realized on single chip [73]. Due to parallel operation, it was possible to digitize and time-stamp the signal at the same time. With an addition of a prescaler to TDC, it was possible to make a time resolution of 5 ps [74].

2.5.6.1.5. ADC with Current Measurements at Input

Ultra-sensitive ADCs for current measurements were also developed, see Table 2-10.

Table 2-10. ADC for the current measurements

Year & Ref.	Group	Parameters	Comments
2002 [73]	HYPRES	4 nA per Least Significant Bit at 100 MS/s, which corresponds to $4*10^{-17}$ C or 250 e. 1.6 microA as total current.	
now	HYPRES	1 microA full-scale range with SQUID front-end	

2.5.6.2. Digital Signal Processor (DSP)

DSP performs all necessary software data processing and amplifies signal for electronics located at room temperature. Signal processing typically includes: i) down-conversion, ii) channelizing, iii) digital decimation filtering, iv) autocorrelation.

2.5.6.3. Decimation Digital Filter (DDF)

For the case, if precise waveform of the signal is not known, but the frequency range is fixed, a decimation digital filter is applied in order to select the required frequency range. The frequency selection is done by a multiplication of the signal by a window filter and improving the signal-to-noise ratio (SNR) by removing noise power, which is located at the outside the required frequency ranges. In order to have a steep response, several integrating and differentiating filters (so called a cascaded integrator comb – CIC) are applied. The first integration stage is required to restore the signal from differential code and the further ones to perform a signal filtering. Typically, a programmable decimator module is incorporated into filter in order to have possibility to make trade-off between bandwidth and SNR. Such filters are also made in modules for flexible reconfiguration and expansion possibilities. It is also worth to note that DDF doesn't need to have ideal characteristics, because final filtering could be performed at room temperature by corresponding "warm" electronics.

For the setting output sample rate, i.e., signal bandwidth, and, therefore, resolution, it is possible to apply a dynamically programmed prescaler, which sets the level of averaging and the output sample rate. This prescaler is incorporated into the filter and allows permanent re-programming during its operation using external controls. Thus, a single ADC chip is capable of instantly changing its bandwidth within the front-end cut-off frequency to zoon-in or to zoom-out on some particular frequency range.

2.5.6.4. Decerializer (Demultiplexer)

If no data processing after ADC is needed, they can be deserialized with the ratios range of 1:8 to 1:16 and transferred to room temperature electronics. It is worth to note that higher decerializing rate is possible, but this is typically not done, due to increased number of channels, which leads to additional heat load on cryogenic system as well as complications on assembly.

2.5.6.5. Correlation-Based Matched Filter

For the case, if precise waveform of the signal is known, it is possible to use a matched filter, which makes the correlation between measured and expected signals [76, 77]. The digital filtering is done in frequency and time domains simultaneously, so uncorrelated noise is effectively suppressed. Other uncorrelated signals are also decreased. Due to high operation frequencies, the filtering is performed in the original band without down conversion in the baseband, see Figure 2-28 below. In order to have required synchronization between received and expected signals, pair of correlators, which are shifted by 90°, are used (in-phase and quadrature signal). One of the signals will be at the maximum and feedback loop keeps the signal locked. In the digital domain, the phase adjustments are achieved in discrete time steps, corresponding to single clock period. Number of required correlation channels is fixed depending on the symbol constellation of digital modulation. Operation at the 20-40 GHz frequencies was demonstrated.

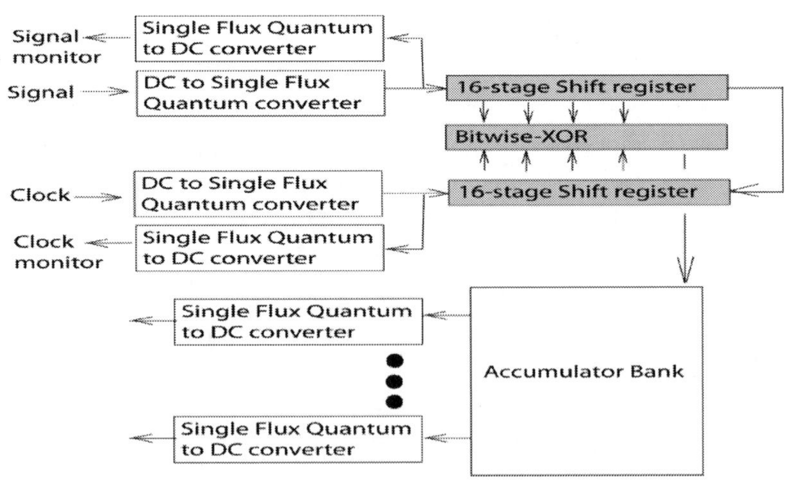

Figure 2-28. Block diagram of 16-bit autocorrelator [76, 77].

Adaptive filtering for baseband non-linear recursive filters for interference rejection was also realized [78].

2.5.6.6. Digital Channelizer

It receives high stream of data from the ADC at the clock rate, 40-50 GHz, performs down conversion, separates the signal auf I-Q components, see Figure 2-29 below. It also typically includes Digital Decimation Filter. Output data are delivered at the decimated clock rate or at higher rates due to parallel-serial conversion stage.

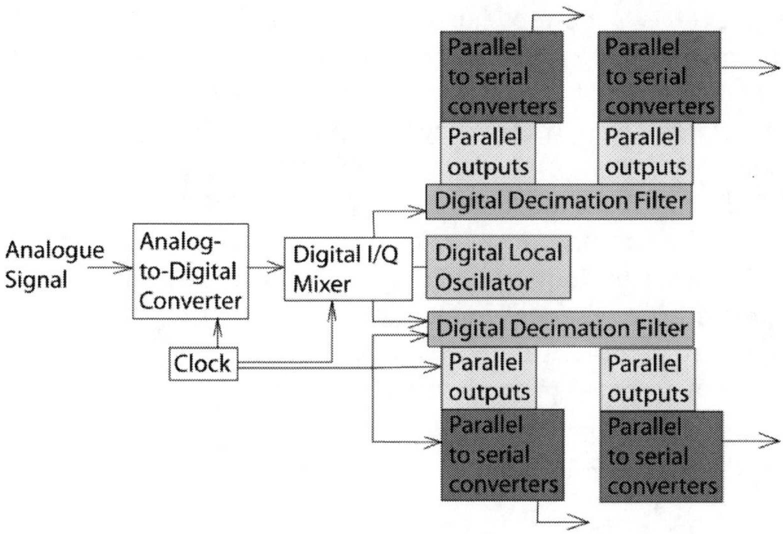

Figure 2-29. Simplified block diagram of a digital channelizer with an in-phase (I) and quadrature (Q) digital mixer and two decimation filter [56]. It is possible to have a version without down-conversion, so the I and Q component are separated and supplied to filters at half of the clock rate. The parallel-serial (P/S) converters (if needed) allow to increase the output rate per single channel.

2.5.6.7. Clock

An on-chip clock technology includes four main components: i) a stable, low-jitter master clock generator, ii) a clock distribution network, iii) a clock decimator and selector to generate and select a sub-harmonic of the master clock frequency, and iv) phase-locked loop (PLL) for synchronization the on-chip master clock with an external (reference) clock to insure the long-term phase stability.

For the practical applications with requirements for time-interleaved SFQ clock signals of complementary phase, two-phase 50 GHz clock sources using Long Josephson Junctions have been developed. Operated in the resonance mode, such junctions have high quality factors in the range of 10^5-10^6. For I and Q components of signals, two clocks with 90° phase shifts are foreseen. A common clock controller with clock distribution network is designed to ensure synchronization between the I and Q data streams. Recently firm HYPRES Inc. also included 1000 GHz clock in their portfolio.

Typically, two physical methods of signal generation are applied [79-82]. One is a flow of fluxon (or a magnetic flux quanta) in an underdampted Long Josephson Junction (LJJ) along the junction between two ends and emitting flux quanta at ends (4π phase change occurs at ends). Due to low damping, such electromagnetic wave has a properties of soliton, and resonance conditions for standing wave in such LJJ occur. If a magnetic field is applied to LJJ, the fluxons are introduced at one end accelerated to the other end. This is another method, which is called as flux-flow mode. In this mode, it is possible to tune frequency by magnetic field, and intensity by bias current of LJJ. Two geometries of LJJ are typically considered, the linear and annular ones. It is also possible to design resonators with high quality factors up to 10^6. The distribution of signals to other components is done over long strip lines (also called “tails”).

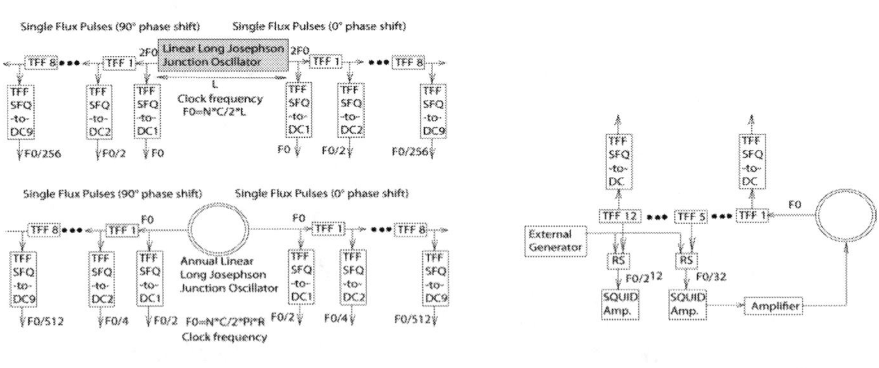

Figure 2-30. (left) linear and annular two-phase clock generators (for I and Q components); and (right) measurement scheme for phase-lock measurements [80, 81]. TFF and RS are toggle and set-reset flip-flops, respectively. SFQ notation means Single Flux Quantum.

Figure 2-30 (left) shows simplified scheme of clock generator with frequency dividers. The signal distribution lines to other components are not shown.

In order to have synchronized operation of chip clock ("internal" one) with external one, the phase-lock loop (PLL) was integrated on chip, see Figure 2-30, right [80]. The circuit on chip includes a binary counters (chain of T flip-flops) with monitoring output, as well as an array of RS latches (RS flip-flops), which are synchronized with external room temperature reference clock. Each RS latch works as a phase detector with dc voltage proportional to the phase difference between decimated output from this annular LJJ and external reference clock. The digital outputs for monitors and RS latches were obtained using TFF-type (T flip-flops-type). SFQ/DC converters transform SFQ pulses for transitions of a voltage waveform of around 250 μV amplitude, which was subsequently amplified in a phase-locked loop to at least 2 mV by on-chip SQUID amplifiers. Final amplifier (before attenuator and loop filter, see Figure 2-30, right) is foreseen to increase the ability of PLL to handle a current load by broadening the phase-locked current range on the fluxon step [80]. Using this built-in PLL on chip with external clock, it was possible to stabilize the 39.7 GHz chip clock to a Δf_{final}=24 Hz linewidth (external clock was 2.48 GHz and 16th harmonic of internal clock was used for PLL) and cycle-to-cycle jitter of phase-locked clock source was measured to be 9 fs at frequency of 39 GHz. It is worth to note that quality factor of this annular LJJ was 4.8×10^6, which corresponds to the frequency bandwidth of $\Delta f=39.7\cdot10^9/4.8*10^6 = 8.3\times10^3$ Hz. So, an application of PLL narrows the frequency span by factor $\Delta f/\Delta f_{final} = 8.3\times10^3/24 = 346$. Supplementary measurements also showed the phase-locked range of around 25 kHz (this range could be slightly changed by changing the supply current of LJJ).

Other measurements of LJJ at frequencies up to 30 GHz showed possibility to apply "injection locking" for clock synchronization. Injection of an external RF signal near natural resonant frequency of Josephson Junction created a vertical voltage-step (also called "Shapiro steps") at the external signal frequency [79].

In 2000, PLL operation at 12 MHz external and 50 GHz chip clocks was also demonstrated [83]. The feedback bandwidth was 50 MHz, phase stability better than one part in 10^3 radians at the reference frequency, and tracking range of ± 1.5% or 49.30-50.70 GHz. Voltage-controlled oscillator, high speed digital frequency divider, phase detector and feedback loop filter (integrator) were designed on-chip.

2.5.6.8. DAC

RSFQ dual binary-type digital-analog converter is based on combination of binary digital-to-frequency converters (DFC) and binary summing SFQ arrays. Each SFQ pulse train from DFCs generates an output voltage across the corresponding seral stage of the SFQ summing array. The final sum of the induced voltages from all stages appears at the analog output.

It is worth to note that DAC operating in giga-Hertz region is still a challenge. Recent developments are based on application of Josephson Arbitrary Waveform Synthesizer (JAWS) done at NIST (National Institute of Standards and Technology, USA) [84-89] and could be summarized as follows:

- Lumped element RF JAWS with 102 Josephson Junctions: 1 GHz sine wave synthesized over 10 MHz bandwidth. The power is -71 dBm, and SFDR is 66 dBc, 10 Hz resolution bandwidth.
- Diplexer RF JAWS circuit with 1440 Josephson Junctions: 1 GHz sine wave synthesized over 10 MHz bandwidth. The power is -49 dBm, and SFDR is 79 dBc (input digital stream was delivered at 4.8 GHz rate).
- SFQ JAWS with DC-to-SFQ converter: 4 GHz sine wave synthesized over 10 MHz bandwidth. The power is -74 dBm, and SFDR is 70 dBc.

2.5.6.9. Direct Digital Synthesizer (DDS) for Digital-RF Transmitter

Digital-to-analog Converters (DAC) converts signal to RF band, which is further supplied to high power module for final amplification. A dynamic digital equalizer, which is typically integrated within DAC, is used for compensation of nonlinearities in high power amplifiers. Digital addition of different channels with different carries frequencies leads to a one broadband signal and as consequence to only one high power amplifier. Additional advantage of high frequency DAC/ADC and processing unit is that RF equalizer (also called as predistorter) can correct instantaneous, signal-dependent fluctuations of the HPA transfer function in picosecond time range. So, superconducting electronics will perform RF predistortion in the digital domain.

DDS consists of several components, i.e., digital interpolation filter (DIF), digital up-converter, a digital sigma-delta modulator (sometimes known as digital encoder) [90], dynamic digital equalizer (also known as phase rotator) and DAC. First and second order DIF circuits to perform digital interpolation, accurate up-conversion of baseband data to RF (125 MS/s to 2 GS/s) followed by digital filtering using a Hogenauer Cascaded Integrator Comb filter have been developed [91].

Figure 2-31 shows the DDD block diagram and waveform generation scheme [91]. Block diagram of digital predistortion filter is presented in Figure 2-32.

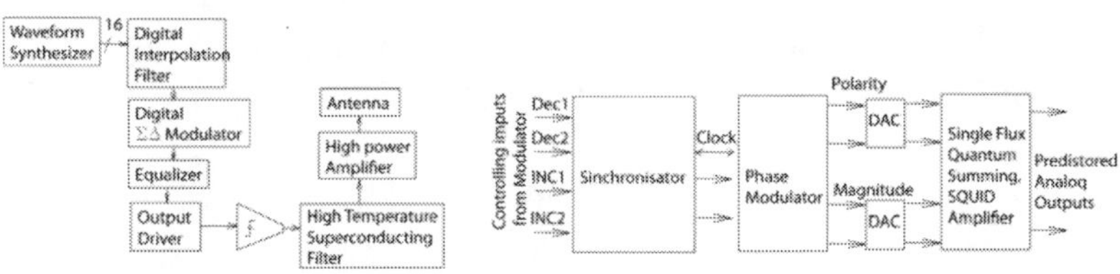

Figure 2-31. Direct Digital Synthesizer block diagram (left) and waveform generation scheme (right) [91].

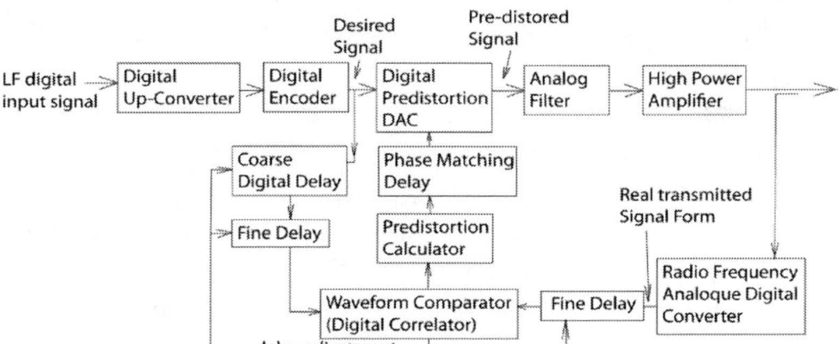

Figure 2-32. DDS predistortion scheme [59, 60] (also noted as “equalizer” in the previous Figure).

The goal of pre-distorter is to linearize the output of High Power Amplifier (HPA). The linearization is typically performed by two methods, i.e., feed-forward distortion compensation and digital pre-distortion.

- Baseband pre-distorters rely on quite complicated digital signal processing algorithms working on the demodulated low-frequency baseband digital waveform with the main goal to attempt to compensate for the amplifiers non-linear gain and phase characteristics. These indirect methods involve either mapping an input in-phase and quadrature signal vector into an output signal vector or multiplying the signal with a level-dependent complex gain. It is worth to note that this method is quite complex and deliver moderate improvements.
- Analogue feed-back compensation in baseband is possible only for slowly changing signals. For that reason, only digital feed-back option operating in required RF range is the only reasonable option for high changing signals, and therefore RSFQ logic must be applied for many cases.

Figure 2-32 (bottom) shows component of predistortion scheme (sometimes called "equalizer") [59, 60]. After digital up-conversion, the multi-bit digital RF waveform is encoded into a single oversample bit-stream, e.g., delta-sigma code, in digital encoder (also called "modulator"). This code is suppled to ΣΔ-DAC, and analog signal is further amplified by high power amplifier. Predistortion involves in adding or substracting of single flux quantum (SFQ) pulses in the sigma-delta code flow. For example, for ΣΔ-DAC, these impulses are added to or subtracted from integrating loop, which is coupled to DAC. For the feed-back loop, signal after high-power amplifier is measured and converted to SFQ pulses in ADC. So, it is possible to compare sample-by-sample (or with other words, "SFQ pulse by SFQ pulse") of the original transmit signal with distorted one. This is done in waveform comparator (also known as digital cross-correlator), which is working on exclusive-OR gated, which is followed by an accumulator to store signal over some time period. Coarse and fine delays allow matching within a fractional clock period by adjusting of the bias of active

Josephson Transmission Lines (JTLs). Next step is calculation of amount pulses to be added to or subtracted from original form. This is done in the predistortion calculator. The final step is a fixing phase matching delay, which is needed for proper phase matching of feedback and original signals.

One of the first superconducting DDS was designed by groups of TRW Space & Electronics Group and Sierra Monolithics in 1997 [85]. It included 12-stage Incremental Phase Accumulator, a 10-bit Sine-Function Generator (SineROM) and 8-bit Digital-to-Analog Converter. The demonstrated parameters could be summarized as follows: instantaneous bandwidth – 500 MHz, frequency resolution 500 kHz, spurious content – -30 dB, clock rate 2 GHz, operation frequency – 1 GHz, and output power from DAC – -60 dBm.

2.5.6.10. Digital-RF (Asynchronous) Switch Matrix (Non-Blocking and Multicasting)

For multi-band communication system with N transceiver and M antennas, the digital asynchronous switch matrix was developed. It has the following requirements [93]:

- Switch must route each of M inputs streams to one or more output terminals, or with other words more generally, multiple input terminals can be connected to one output, if data collision conditions are noticed.
- Switch must support multiple input data rates, because ADC may not have the same clock frequency.
- Operation at a rate up to 40 GHz.

2.5.6.11. Data Transfer

Data transfer from low temperature electronics to one working at ambient temperatures is surely a challenge and is one of the reasons for handicapping a wide application of cold digital electronics [94-107]. This is related to the contradicting requirements of low heat loads, which requires uses of long cables with minimal cross-sections areas, and high data transfer, which implies short coaxial cables with metal conduit. Moreover, other requirements, like cable flexibility for adapting of material shrinkage at low temperatures, thermal anchoring at intermediate temperature or access easiness for cable repair or installation of additional cables, lead to enormous engineering challenges for cryogenic engineers and several design compromises are typically done.

Over the last 20 years, the HYPES Inc. company made many design changes and developed several methods for the data transfer as well as made significant progress in the hardware development. Different data transfer rates per single channel (i.e., "cable") were realized, e.g., i) SQUID-stack "Dodo" – 6 Gbps, ii) SQUID-stack "BlueBird" – 14 Gbps, iii) SQUID-stack "Ostrich" – 30 Gbps, and iv) differential SFQ/DC – 40 Gbps [94], see Table 2-11, and Figure 2-33. Depending on each application, it is possible to choose "old" output drivers with larger number of channels, or "new" one but with smaller channel numbers. For example, Field Programmable Gate Array (FPGA) is typically connected to the interface amplifier with 32 differential (LVDS) channels, which stream the digitized data into the FPGA using double rate mode (DDR). So, a set of 32 consecutive bits is arriving every $f_{clock}/64$ cycle at the FPGA input. Such systems have been demonstrated and proven to support aggregate data rates up to 36 Gbps (e.g., lowpass phase modulation-demodulation Δ-ADC clocked at 36 GHz). Other designs, which are routinely in operation, supported data transfer per single channel 6-14 Gbps and being able to transfer 120 Gbps totally to single FPGA. Further design, which is based on the SQUID-stack driver (SQUID-stack driver is used as an amplification stage after RSFQ signals and typically delivers signals in the range 1-3 mV) could deliver the data rate per single channel up to 30 Gbps.

Table 2-11. Output drivers for superconducting digital electronics [94]

Output driver	Data rate per single channel (Gbps)	Swing @ 50 Ω (mVpp)	FOM (fJ/bit)
SQUID-stack "Dodo"	6	5	16.5
SQUID-stack "BlueBird"	14	5	9.6
SQUID-stack "Ostrich"	30	5	5.3
Differential SFQ/DC	40	1.3	0.3

On-chip measurement of data transfer with output driver “Ostrich” showed that rates up to 30 Gbps are possible. The data transfer to room temperature was limited to 18-20 Gbps due to interface amplifier.

Channel synchronization (also known as bonding) was successfully demonstrated up to 40 GHz, which corresponds to 10 Gbps per channel in 1:4 deserializer configuration [108].

Presently, output driver “BlueBird” is widely applied and is a “workhorse” due to the mature of technology.

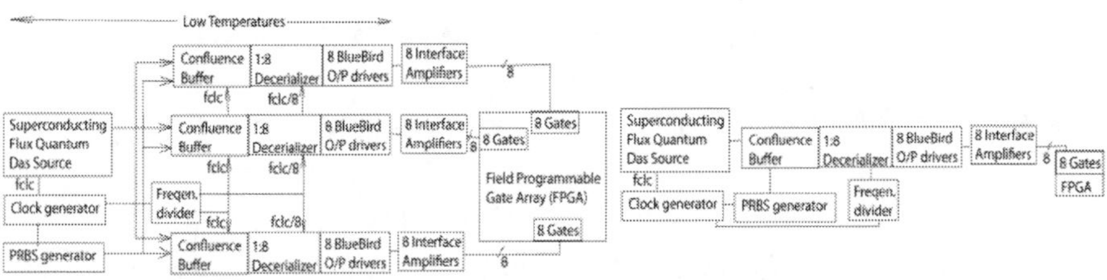

Figure 2-33. Example of BlueBird data link [94]. For this example, pseudo-random bit sequence (PRBS-7) generator (left figure) is used to synchronize the banks of 8 parallel channels (deserializer ratio is 8). The GTY data receivers on the Xilinx FPGA are asynchronous and by using three decerializer banks, the 120 Gbps total rate is achieved (limitation was on PRBS, which did not allowed higher rates. With one decerializer bank (right), the maximum data transfer was 64 Gbps.

In order to realize such high transfer rates, two concepts are developed, i.e., with and without cold intermediate amplifier at 40-80K temperature rate, see Figure 2-34, Figure 2-35 below [104].

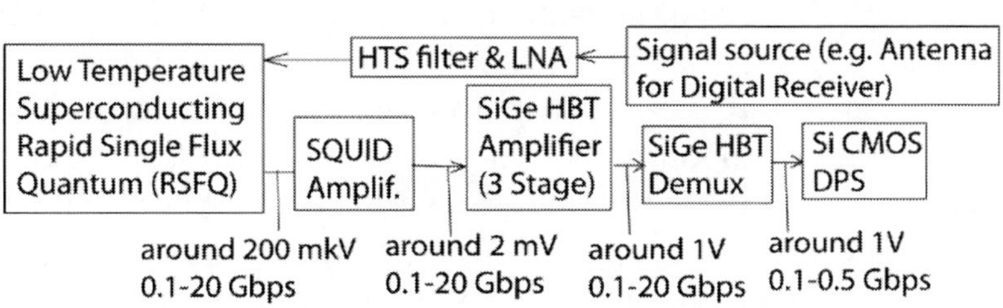

Figure 2-34. Architecture of superconducting digital system [104].

For amplification at room temperature, double stage amplifier with transimpedance circuit as well as Schmidt trigger were used. Amplifier bandwidth increase was approximately linear with temperature, i.e., around 6 GHz at room temperature and 11 GHz at 77K.

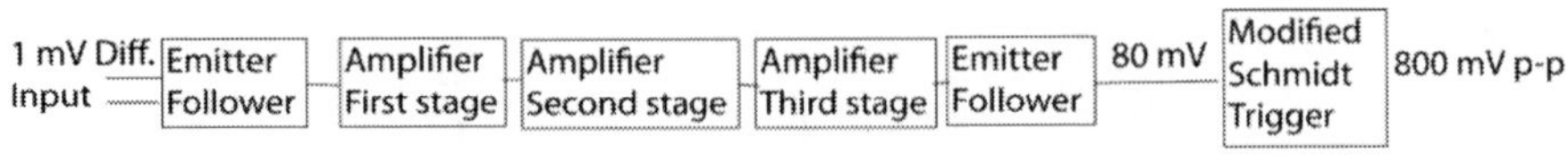

Figure 2-35. A SiGe heterojunction bipolar transistors (HBT) amplifier with emitter follower buffers and modified Schmidt trigger. Data are converted from mV-level to V-level single-ended ECL signal [104].

Another example of wide-band SiGe amplifier is presented in reference [109].

Recently, HYPRES Inc. Company made a progress in the development of multi-stage SiGe preamplifiers, which are together with coaxial cables often limit data transfer from low to room temperatures [94]. A new compact 20-channel datalink amplifier unit was developed and built. It consists of two 10 channel circuit boards. Each channel contains a low noise linear gain stage (32 dB), and a discriminator that outputs standard CML logic signal. RF connectors allow for direct plugin of this 20-channel amplifier into the 20-coax RF vacuum feedthrough (G3PO-type) of the cryogenic system. This way of connecting allows eliminating input cables that were typically 0.6 m long and were potentially susceptible to EMI interference. Each of the 10-channel amplifier boards includes a commercial interface for a ribbon of differential twin-coaxial cables for direct interface to an FPGA board. The whole unit, including the metal encasing, has volume of $55 \times 30 \times 58$ mm^3. The power consumption is 4 W, operating from a single 3.3 V power supply. Together with the Bluebird output driver and the stripline BeCu-Cu-BeCu ribbon cable version 3, this interface amplifier has proven data transport from cryogenic chip to FPGA at 8 Gbps per channel. It worth to note, that this stripline ribbon cable caused some cross-talk between cables and should not be applied for high-speed data transfer. It was possible to operate a single channel with data transfer rate to around 14-18 Gbps. The gain of double stage SiGe amplifiers at room temperature as approximately 30 dB in the frequency range of 2 till 10 GHz, while for the single-stage one were 12-14 dB (operation at room temperature) and 14-15 (operation at cryogenic temperatures, 41K) [94].

Table 2-12. Other firms or research institutions, which have experience with RSFQ components

Name	Development
Nagoja University, Kinki University, Japan	Prototypes of ADC and digital filter were developed.
Superconductivity Research Laboratory of International Superconductivity Technology Center (ISTEC)	Advanced development of flash ADC and digital filters
Physics Department, SUNY at Stony Brook, USA	Development circuits and manufacturing of some simplified chips. Group is not active in this field any more.
Northrop Grumman Science and Technology Center	ADC developments. Not more active in this field.
Institut für Schicht- und Ionentechnik, KFA, Jülich, Physik-Department E 10, TU München.	Comparator development. Not more active in this field.
National Institute of Advanced Industrial Science and Technology, Japan	Memory, RSFQ computing, ADC
D-Wave System Inc.	Memory
MIT Lincoln Laboratory Systems Inc.	Memory, logics
CEA-INAC, LETI Grenoble, France	Logic based on NbN-TaN-NbN junctions
Microtechnology and Nanoscience, Chalmers University of Technology, Sweden	Development of digital signal processor for telecommunication. Chip is from HYPRES.
Quantum Electrodynamics Division, National Institute of Standarts and Technology, USA	DAC.
Saitama University, University of Tokyo, Cadence Design System Inc.	Digital DROS
National Institute of Advanced Industrial Science and Technology and Yokohama National University	Digital SQUID

Besides HYPRES company, there are other companies or scientific institutions, which have some experience in design, development or application of components working of RSFQ logic family, see Table 2-12 above.

2.5.7. Future Trends and Area of Possible Research Activities

Though present design of the ADR system is in very advanced stage and finds applications, see Appendixes A and B, it is still possible to make further developments, i.e.:

- Application of other low-Tc materials, which have higher operation frequencies: it is possible to use other materials with higher superconducting/normal conducting transition temperatures as Nb one.
- Other high-Tc systems, e.g., YBCO material, could be applied.

Future activities toward further development of radar and communication systems could be related to the enhancement of signal detection, e.g.:

- Compact devices with normal or superconductor (low- or high-Tc materials) antenna. Applications of tunable sc filter and LNA before ADC could further improve the sensitivity.
- Develop system with tunable (sc) antenna and tunable (sc) impedance matching.

2.6. Superconductor Antennas

Although, there was quite significant and interesting work done on antennas, which were constructed from superconducting materials, non-of them has been applied in commercial receivers or transmitter. This is related to the fact, that the overall benefit is very limited and the only chances to come to the market are:

- to design whole complete receiving front end, i.e., sc antenna, sc filter, LNA, on components, which are cooled to low temperatures; or

- for electrically small antannas with quite significant losses due to miniaturization, active impedance matching or active antenna tuning could be very beneficial.

However, for the sake of completeness, a short review on sc antenna is given here.

2.6.1. General

It is also possible to consider differences between normal- and superconducting "classical" antennas, like dipols, slots, spirals, reflectors, etc. The performance of nearly all normal conducting antennas is defined by size and shape in the free space wavelength, i.e., on the external fields. In case of superconducting material, superconductivity has some minor effect on external field, and therefore, antenna "shapes and sizes" are practically not changed (or slightly reduced). However, the internal fields inside the superconducting material are absent; and, therefore, efficiency is improved, and it is possible to increase the efficiency of impedance matching network. For the calculations and practical realization of superconducting antenna, it is important to respect some limitations or additional effects, e.g., thickness of superconducting material should be larger than London penetration length λ; for non-zero frequencies, the superconducting electrons have kinetic energy, which produces delay in electric and magnetic field along the surface, and therefore, "kinetic" inductance must be taken into account; or surface resistance and reactance have different frequency dependencies in comparison to ones of normal conducting materials [114].

In order to handle high powers at antenna or filters, the edge free modes are usually applied, otherwise too high magnetic field strength leads to transition of superconducting to normal conducting state of superconductor. For circular geometry, TM_{mn0} or TM_{0n0} modes are particular suitable for such applications.

2.6.2. Loop and Dipole Antenna

Historically these types were the first investigated. Figure 2-36 shows the simplified view of short dipole and loop antennas.

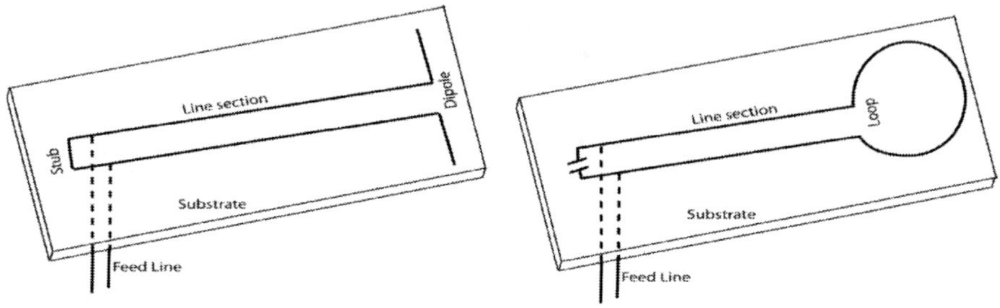

Figure 2-36. Small dipole and loop antennas with corresponding matching networks [112, 114].

Table 2-13 summarizes some measurement results on small dipole and loop antennas.

Table 2-13. Short summary on gain values of superconducting antenna in comparison to Au or Cu ones

N	Type	Lengths: l/l1/l2	Frequency	Improvement/Gain	Reference
1	dipole	20/77/2.1mm	500-580 MHz	Relative field strength: 12dB in comparison to Cu at 300K or 6 dB in comparison to Cu at 77K	[110, 111]
2	loop	Ø50/17/5mm	420-460 MHz	Relative field strength: 9dB in comparison to Cu at 300K or 5 dB in comparison to Cu at 77K	[112, 111]
3	helix	20mm, two loops	350-550 MHz	Gain in broadside direction: 9dB in comparison to Cu at 300K or 5.8 dB in comparison to Cu at 77K	[113]
4	end-fire	λ/2, two elements	1 GHz	Gain: 3-8dB in comparison to Cu at 300K or 2-3 dB in comparison to Cu at 77K	[113, 115]
5	loop	Ø17.5/_/1.5mm	300 MHz	Gain: 6.7dB in comparison to Au at 77K	[116]

N	Type	Lengths: l/l1/l2	Frequency	Improvement/Gain	Reference
6	loop	Ø18-40/_/1.5mm	250-500 MHz	Gain: 3.5-6dB in comparison to Cu at 77K	[117]
7	loop	Ø3.5/-	663 MHz	Gain: 24 dB (total) in comparison to Cu at 300K (18.9 due to antenna efficiency improvement and 5.1 dB due to reduction of matching losses) $T_{antenna}$=4K	[118]
8	helix	Ø10mm, 3mm pitch, 2.7 turns	897 MHz	Gain: 3 dB in comparison to Cu at 77K	[119]
9	helix	Ø10-12mm, single or two ant.	500, 900 MHz	Gain: a), two antennas, 7.8 dB in comparison to Cu at 300K or 6.2 dB in comparison to Cu at 77K, b) single, 9.6 dB in comparison to Cu at 300K or 5.4 dB in comparison to Cu at 77K	[120]

It is worth to note that the most significant contribution to the efficiency increase was caused by impedance matching effect. Matching a radiation resistance of an ohm or less to 50 Ohm lines incurs significant losses, similar to matching a reactance of thousands of ohm [121, 122]. For example, for a transmission line with loss *L*, which is connected to a load with VSWR>1, the apparent losses are $La = [(VSWR+1)^2 * L^2 - (VSWR-1)^2]/(4 * VSWR * L)$. So, small intrinsic losses L could be significantly increased due to large VSWR.

2.6.3. Microstrip Antenna

Figure 2-37 below shows the typical microstrip antennas.

Table 2-14 summarizes some experimental results on the microstrip antennas.

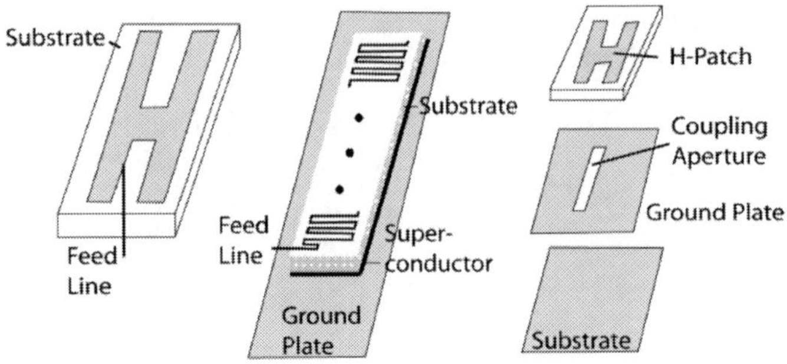

Figure 2-37. H-Patch and meander patch antennas [123-127].

Table 2-14. Microstrip antennas

N	Frequency	Improvement/Gain	Reference
1	2.45 GHz	Radiation efficiency: 35-65% in comparison to 1-6% for Cu at 77K	[123]
2	2.3 GHz	Gain is 5.5 dB higher than for Cu (300K), and efficiency is increased to 18.2% (for Cu-4.6%)	[124]
3	38 GHz	Radiation efficiency: 1.2 dB in comparison to Ag at 80K	[125]
4	20 GHz	Efficiency is increased to 90% (for Ag-78% at 77 K)	[126]
5	5 GHz	Gain is 4 dB higher than for Au (80K)	[127]

2.6.4. Array Antennas

Figure 2-38 shows several possible array structures manufactured from high-Tc superconductors.

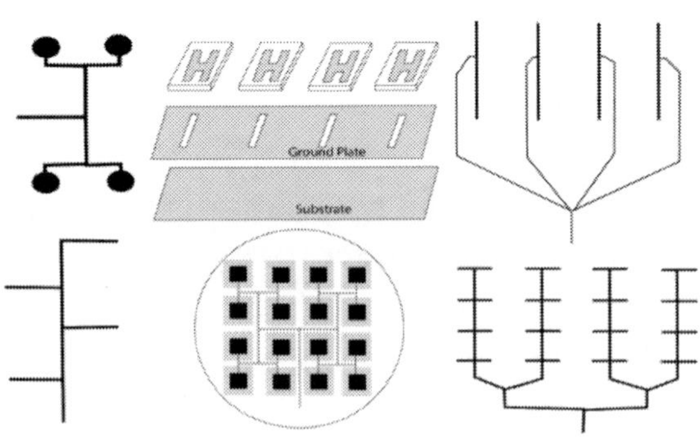

Figure 2-38. Array antennas [128-141].

Table 2-15 summarizes some experimental results on the microstrip array and end fire (number "3") antennas.

Table 2-15. Microstrip array antennas

N	Frequency	Improvement/Gain	Reference
1	26-29 GHz	Gain: 9-10 dB in comparison to Au at 300K (see Figure 2-38, top, left)	[128]
2	3-12 GHz	Three array demonstrated multiband self-diplexing multifrequency-enhanced bandwidth and multifrequency beam forming (see Figure 2-38, top, center)	[129]
3	1 GHz	Gain: 3-8dB in comparison to Cu at 300K or 2-3 dB in comparison to Cu at 77K	[115]
4	1 GHz	Gain: 4-6B in comparison to Ag at 300K or 3-4 dB in comparison to Ag at 77K (see Figure 2-38, bottom, left)	[130]
5	1 GHz	Gain: 3-6B in comparison to Ag at 77K (see Figure 2-38, top, right)	[131]
6	1 GHz	Gain: 4B in comparison to Ag at 300K or 3 dB in comparison to Ag at 77K (see Figure 2-38, bottom, right)	[132]
7	12 GHz	Increase in the gain, though exact values are not given (see Figure 2-38, bottom, center)	[133]
8	12-20 GHz	16-Element Array was designed	[134]
9	12 GHz	Gain: 3.6 dB in comparison to Cu at 300K	[135]
10	20 GHz	Gain: 2.2 dB in comparison to Cu at 300K	[136]
11	31 GHz	Gain: 3.5 dB in comparison to Au at 300K	[179]
12	5.8 GHz	Gain is 8 dB higher than for Cu (300K), and efficiency is increased by 69.2% in comparison to Cu	[137]
13	11 GHz	Gain: 1.54 dB in comparison to Au at 300K	[138]
14	11 GHz	Gain: 2.5-8 dB in comparison to Au at 300K	[139]
15	30 GHz	Efficiency: 4.7 dB in comparison to Ag at 300K or 2 dB in comparison to Ag at 77K. Note: antenna was from normal conducting material, but feeding network – from superconducting.	[140]
16	10-11 GHz	Gain: 3.5 dB in comparison to Cu at 300K	[141]

2.6.5. Trends and Future Activities

At the present, quite limited work is going on related to the sc antennas. Main investigations are related to the microstrip antenna with some more attention to the loop one.

It is worth to note that over the last 10 years, two trends have been noticed (particularly for warm antennas with the small geometries):

i) reduction of losses in the matching network, and ii) active antenna tuning for different frequency ranges.

For the future activities on sc antennas, these two improvement possibilities are promising and should be investigated, i.e.:

- active impedance matching between antenna and other components,
- active tuning of superconducting antenna, e.g., by aperture tuning with switch.

2.7. Antenna Based On SQUID/SQIF Elements

In the recent years, several groups tried to develop superconducting antenna based on operation of SQUID/SQIF/bi-SQUID blocks. Such antennas belong to the class of electrically small active one, have high sensitivity and are broadband. Typically, due to sufficient high signal at the outlet, they are directly connected to LNA operating at the room temperature. These antennas are typically classified as ones working at temperatures 4-8K and made from low-Tc superconducting materials, as well as ones at 40-80K, which are manufactured from high-Tc superconductors.

Besides application for radar and communication systems, other application areas are:

- detectors for ore content and mines (LANDTEM, GETMAG [142]) [142, 288, 289] and undetonated explosives on see floors,
- explosive components deep in body (UXOMAG, [142]), [288, 289],
- superconducting detector-based THz imager [142, 145, 146, 288, 289],
- frequency-tuneable gigahertz heterodyne oscillators and mixers [142, 143, 144, 288, 289],
- MMIC down-converter and receiver front-ends up to Ka-band (gain >40 dB and noise figure <0.5 dB) [288, 289].

Main advantages of SQIF applications were related to improvement possibilities of the signal-to-noise ratio (and as consequence – dynamic range), as well as the impedance matching for efficient signal transfer and insensitivity to the external (disturbing) magnetic fields. Moreover, the output voltages are quite high, >10mW, so inexpensive room temperature electronics can handle the received signals, as well as a very large bandwidth is also realizable.

It is worth to note that applications of SQIF antennas were considered for several military programs, e.g.:

- USA [67, 147]: Tactical SIGINT Technology Program N66001-08-D-0154, SPAWAR SBIR contracts N00039-08-C-0024, N66001-09-R-0073, ONR (Code 30), the SPAWAR internal research funding (S&T) program and HYPRES R&D, see also Appendix B,
- Australia [148]: Supported by Department of Defense Science and Technology Group (DSTG), through CTG programme.

In this chapter, an overall activity done over the last 10-15 years on SQIF/bi-SQUID antennas is overviewed. We start with a very short theoretical background of SQIF/bi-SQUID antennas, followed by short discussion of some prototypes built and published in the opening literature sources.

It is worth to mention that due to limited market, very few firms or research institutions are active in this field. Therefore, in some cases, it is quite difficult to obtain detailed design information on each particular antennas.

2.7.1. General

Operation of the Superconducting Quantum Interference Filter (SQIF) and bi-SQUID (SQUID is Superconducting QUantum Interference Device) is based on operation of dc-SQUID, i.e., SQUID with two Josephson Junctions (JJ).

Bi-SQUID is obtained by adding an additional JJ, and linearity of Voltage-to-Flux characteristics is significantly improved. Such bi-

SQUIDs could be connected by groups in parallel and/or series to form arrays of bi-SQUIDs.

Original design of SQIF was proposed as several JJs connected in parallel and with different loop areas between each two JJs, see Figure 2-39.

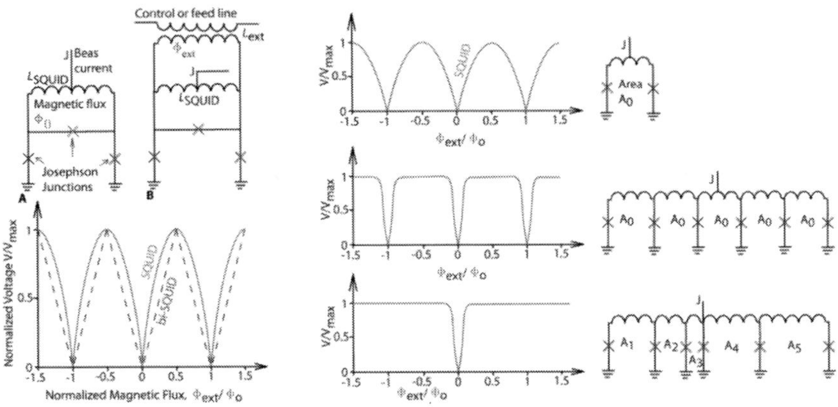

Figure 2-39. Bi-SQUID (top left "A"), equivalent circuit of bi-SQUID with additional transformer loop coupled to a control stripline for magnetic flux applications (top left "B") [149], and voltage response of dc-SQUID (solid line) and bi-SQUID (dashed line) (bottom left) [149], voltage response in units of Ic□R (JJ critical current multiplied by JJ resistance) versus external magnetic flux for dc-SQUID (top right), periodic 1-D array (middle, right), and random array (bottom right) of SQIFs [150].

Main application advantages of SQIF/bi-SQUID arrays could be summarized as following:

- Insensitive to manufacturing variations of single JJ.
- Only one minimum is presented, and therefore an absolute measurement of magnetic field is possible.
- Robustness to external noise, so no external shielding of magnetic field is needed. It is worth to note that this effect was noticed also for large arrays (100 or more) of SQUIDs in 70th, though due to manufacturing uncertainties of single JJs, it was very difficult to produce such large arrays.
- It is possible to increase the absolute value of the voltage difference at minimum by increasing the number of SQIF/bi-

SQUIDs. Thus, a larger dynamic range could be obtained as well as impedance matching is facilitated.

- It is possible to improve the *signal-to-noise* ratio by increasing of SQUID/SQIF/bi-SQUID number.

Besides 1D-parallel arrangement of JJ for SQIF, it is also possible to arrange them in series. It is also possible to have a mixing arrangement of parallel/serial SQIF. In order to have a clear representation of "array structure" of SQIF/bi-SQUID, the following notation is used [151-155], see Figure 2-39 and Figure 2-40:

- Differential Quantum Cell (DQC), which consists of two parallel Josephson Junction array (DQC shoulders), see Figure 2-40 (left). Typically, each of two parallel arrays consist of 10-20 parallel JJs.
- Each DQC is connected in series; see Figure 2-40, where "Sn" describes serially connected DQC with number of *Ks*. These serially connected DQC are named as "blocks."
- Several blocks (each consisting of Ks serially connected DQC) are connected in parallel. Number of blocks is *Kp*. If differential measurements are foreseen, these parallel blocks belong to one "arm" (or also sometimes called "shoulder").
- Two "arms" (or "shoulders") are used for the differential measurements over resistor R_L.

Naming of Superconducting Quantum Cells (SQC) is used to describe basic blocks of Superconducting Quantum Array (SQA). At the present time, SQC describe two types of cells, i.e., bi-SQUID and DQC. By parallel/serial (Ks/Kp) arrangement of SQC, it is possible to obtain different SQAs. By increasing number of SQC it is possible to reduce noise level (and consequently – dynamic range is increased), and by different combination of Ks/Kp blocks or cells within one shoulder it is possible to make impedance matching.

It is worth to note that for some "simple" structure with number of JJ up to 100, "array" naming could be used for serial arrangement of few parallel SQIFs (these parallel SQIF have typically 10-20 JJs connected in parallel).

In some cases, it is also possible to make modulation of parallel/serial arrangement of JJ in order to obtain linear characteristics of array [156-157].

In the following discussion, under naming SQUID/SQIF/bi-SQUID antenna, we imply Superconducting Quantum Array, which belongs to the class of active electrically small antennas (ESA).

The directional pattern of ESA is very similar to |cos|-like dependence (the only difference is pattern for α-angle is sometimes slightly inclined).

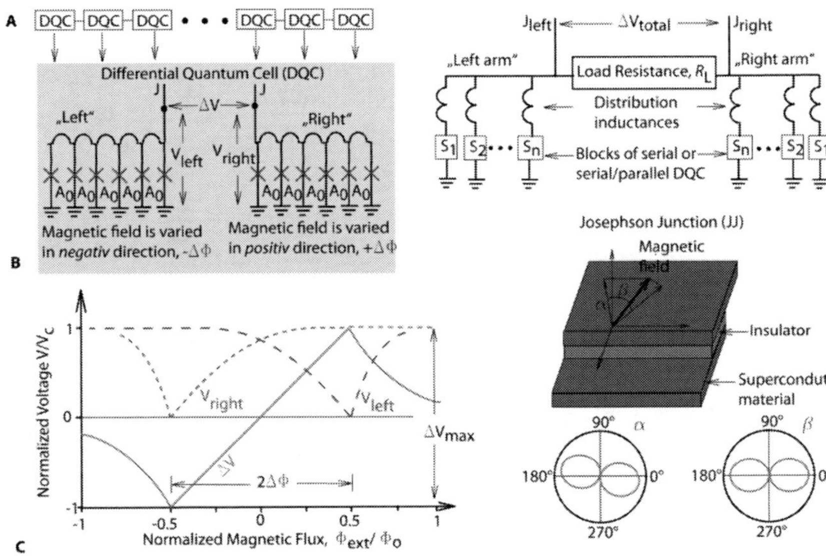

Figure 2-40. Schematic view of 2-D Superconducting Quantum Array consisting of serially connected Quantum Cells, for example of DQC or bi-SQUIDs (top left "a"); DQC comprising of two parallel Josephson Junction Arrays (DQC shoulders) (middle left, "b"); voltage response of DQC (bottom left, "c"). Schematic view of Superconducting Quantum Array (SQA) composed of Kp blocks "S" connected in parallel, each consisting of Ks DQC connected in series (top right); a two-superconductor-plate (film) model of single JJ of DQC (JJ are located between two planes) (middle right); and directional pattern of active electrically small antenna (bottom right) [151].

There is further method for the enhancement of the signal sensitivity, namely "flux concentrator" according to typical naming in SQUID community. This is superconducting washer, which is placed around SQUID (or SQIF) loop. As electromagnetic wave doesn't penetrate the superconducting material, it will be concentrated inside the SQUID loop.

Typically for SQUID, the flux concentration (also called "flux enhancement") is in the range 10-100, and for example for SQUID loop of 1mm^2, washer has 100-200 mm^2 surface area. Similar concept is also done for SQIF/bi-SQUID array. Two versions are possible:

- With transformer loop (one or two): one loop from superconducting material is formed over large areas, so magnetic field from this area will be "concentrated" inside small loops of each Differential Quantum Cells (SQIF/bi-SQUID). Along the perimeter of this large loop, the Differential Quantum Cells are located. This large loop (typically up to couple mm^2) will have smaller loops around each DQC, and magnetic flux of large loop will be "applied" to each DQC.
- Without transformer loop: in this case, the superconducting washer with holes will be located beneath (or sometimes around) each DQC. In some cases, due to fabrication process, DQC with large thick superconducting connecting wires/strips is done, so effective flux concentration by factor 10 is easily obtained.

It is worth to note that transformer-less version of ESA allows higher packing density of DQC and is somehow preferable.

There is another method of "enhancement" of magnetic flux, which is also called "transformer loop" according to classical SQUID notation. It is a loop, made from superconducting material (in former times, practically always from NbTi wires). One end of loop (typically circular form with area 10-10000 mm^2) is exposed to the magnetic field to be measured. After that the magnetic flux is "transported" by superconducting current in twisted pair of sc wires to another smaller loop (but with larger number of turns, typically up to 50), which is located around SQUID. So, magnetic field is "transported" from one place to another and also "concentrated" in smaller area of SQUID loop. From opening literature sources, it is not clear, whether such method has been applied for SQIF/bi-SQUID antenna, because such antennas are already very sensitive.

2.7.2. SQIF/bi-SQUID Antenna Working at 4K Temperature Level

Direct measurements of rf signals in the range of 8-15 GHz by bi-SQUID SQIF showed signal gain in the range of 25-35 dB [181]. A small ring or patch sending antennas in front of SQIF one with the input power or -131 dBm were used for signal transfer. The smallest gain for all frequencies was around 5 dB.

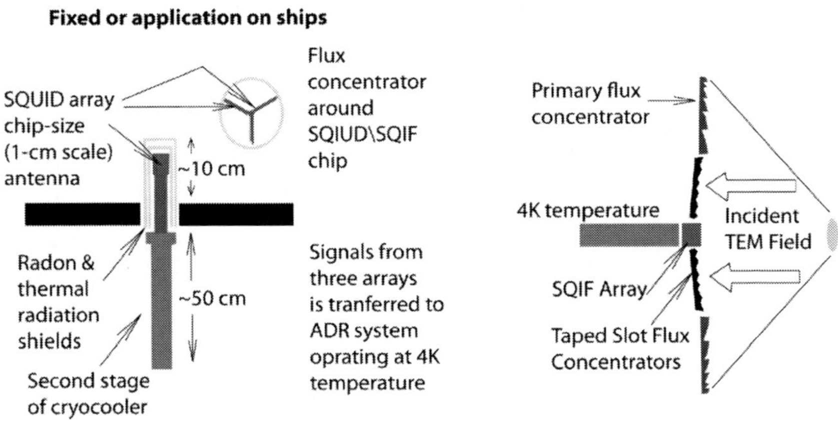

Figure 2-41. An example of application of SQIF antenna working at 4 K temperature level [67]. It is worth to note that according to this figure, after SQIF antenna, ADR-system of HYPRES Inc. company was used.

Other measurements on SQIF arrays in the frequency range 300 kHz - 1 GHz showed excellent performance and gain of around 25 dB [175]. The rf signal was applied through the loop around SQIF array. It is also possible to directly measure rf signals from broadcasting stations by directly exposing SQIF arrays to electromagnetic signals.

Figure 2-41 shows an example of application of 4K SQIF antennas [67]. In this case, superconducting and tapered slot flux concentrators were used. Three chips were vertically arranged and each covered 120°.

Three superconducting quantum arrays, each with 2400 JJs, was manufactured from Nb material. Table 2-16 summarizes some array parameters [158].

The measured Noise Temperature of Superconducting Quantum Arrays could be by 1-2 orders of magnitude higher than the quantum limit [67].

Table 2-16. Parameters of superconducting quantum array

Junction Parameters		
Critical current density, J_c	4500	A/cm^2
Critical current, I_c	80	μA
Resistance, R_n	6.6	Ω
SQUID Dimensions		
Hole width (B)	3	μm
Hole length	3	μm
Effective width	10	μm
Effective length	7.2	μm
SQUID area	72	μm^2
Inductivity of single SQUID	5	pH
$I_c \cdot l/\Phi_0$	0.18	
Cell width	10	μm
Cell length	40.5	μm
SQUID area		
N	30	Columns
M	80	Rows
N□M	2400	JJs
Width	300	μm
Length	3240	μm
Chip Circuit		
# SQAs (only 3 were used)	5	
Sum SQA Widths	1500	μm
SQA length	3240	μm

2.7.3. SQIF/bi-SQUID Antenna Working at 40-80K Temperature Level

Measurements on SQIF antenna, which was constructed from 100 parallel high-Tc SQUIDs and operated in the open loop mode, i.e., voltage output from the SQIF was directly connected to room temperature amplifier, were performed in the frequency range 0-10

GHz and without any magnetic shielding [159]. The cut-off frequency of 10 GHz was set by the inductances of loop array. For the actual measurements, the cut off frequency was limited by the bandwidth of cables and preamplifier. For the case of removed magnetic shielding, an applied rf-signal led to degradation of the minima (unique sharp dip of SQIF). The rf-signals could be easily recovered by a Fourier transformation of the recordered SQIF voltage output [159].

Other measurements showed that SQIF sensor could be used for transferring the modulation of an incident signal with carrier frequency up to several GHz into corresponding rf voltage signal [160]. It could be also used as non-linear mixer from dc to 20 GHz with large dynamic range [160, 161].

Another experiment, where two rf signals up to 20 GHz were mixed in SQIF, which was exposed by static magnetic field produced by circular coil, was also carried out.

Similar to application on the ground station or ships, see Figure 2-41 above, it is possible to mount high-T_c SQIF antenna with single stage cryocooler on small airplane, like “Predator” [67]. In this case low power single stage cryocooler will be sufficient for operation of SQUID antanna at 40-80K temperature level. It is worth to note that after SQIF antenna, no ADR-system of HYPRES Inc. company was used.

It is worth also to mention that significant contribution to the development of large SQIF array was done at SCIRO corporation, Australia [148, 288, 289]. Table 2-17 summarizes some parameters of SQIF array. A 1 cm^2 chip could be mounted on compact single cryocooler with overall length (from cold tip till end of helium storage vessel from Sunpower company) with total weight below 7 kg and overall dimensions 35□18□12 cm. Such system could be portable with battery operation up to 13 hours.

Table 2-17. Parameters of superconducting quantum array

Number of elements	averaged area, (μm^2)	β_L (for Ic =20 μA)	Averaged voltage Vp-p (mV)	Avaraged sensitivity V_B (kV/T)
20,000	280	1.0	1.8	2.85
20,000	30	0.4	6.6	6.35
100,000	30	0.4	24.8	24.8

Table 2-18. Some operational parameters of SQIF/bi-SQUID antennas

N	Material	Number of Cells/Josephson Junctions	Area	Parameters,	Comments	Ref.
1	$YBa_2Cu_3O_7$	105, series connected	10*10mm	$S^{1/2}$=1.2 pT/$Hz^{1/2}$ ΔV=7.5 mV dV/dT=7500 V/T, f=10kHz Estimations: ΔB_{max}=0.0075/7500=1μT For Δf=1GHz, ΔB_{min}=38 nT Dyn. Range =20log(ΔB/ ΔB_{min})=28.4 dB	Effective flux collecting area for a single 20 μm wide junction, is about 350 μm^2, i.e., 25 times larger than the geometrical area. This is attributed to flux focusing effect of the junction's electrodes. For the 105-junction array embedded in the flux focusers, a further increase of the effective area is achieved by a factor 4.2	[162]
2	Nb	29 (parallel)	5*5 mm	$S^{1/2}$≤5 pT/$Hz^{1/2}$ ΔV=67 μV dV/dT=0.2 μV/nT, f=1kHz Estimations: ΔB_{max}=67μV/0.2 μV/nT =335nT For Δf=1GHz, ΔB_{min}=158 nT Dyn. Range =20log(ΔB_{max}/ ΔB_{min})=6.5 dB	A_{totEff}=2717 μm^2. Operation without flux focusing	[163]
3	Nb	29 (series)		$S^{1/2}$≤1.4 pT/$Hz^{1/2}$ ΔV=1350 μV dV/dT=0.72 μV/nT, f=1kHz Estimations: ΔB_{max}=1350μV/0.72 μV/nT =1875nT For Δf=1GHz, ΔB_{min}=44.2 nT Dyn. Range =20log(ΔB_{max}/ ΔB_{min})=32 dB		[163]

Table 2-18. (Continued)

N	Material	Number of Cells/Josephson Junctions	Area	Parameters,	Comments	Ref.
4	$YBa_2Cu_3O_7$	30 (all types)	1.3*0.5mm	$S^{1/2}$=2 (ser), 5(par), 1.3(ser-par) pT/$Hz^{1/2}$ ΔV=440(ser), 52(par), 72(ser-par) μV dV/dT=2100 (ser), 520(par), 320(ser-par) μV/μT,	Operation without flux focusing. Serial, parallel and serial-parallel combination from 30 loops was measured	[164]
5	$YBa_2Cu_3O_7$	30	1.3*0.5mm	$S^{1/2}$=190 (ser), -(par), 70(ser-par) fT/$Hz^{1/2}$ ΔV=390(ser), 16(par), 120(ser-par) μV dV/dT=30 (ser), 3.3(par), 30(ser-par) μV/nT, Estimations: ΔB_{max}=120μV/30 μV/nT =4nT For Δf=1GHz, ΔB_{min}=2.2 nT Dyn. Range =20log(ΔB_{max}/ ΔB_{min})=5.2 dB	Similar to previous case, only pick-up loop (concentrator) Ø8.8mm was added.	[165]
6	Nb	290 (10 rows each with 30 JJ in parallel)	1.2*0.8mm	$S^{1/2}$≤0.5 pT/$Hz^{1/2}$ ΔV=730 μV dV/dT=15 μV/nT, f=1kHz Estimations: ΔB_{max}=730μV/15 μV/nT =49nT For Δf=1GHz, ΔB_{min}=15.8 nT Dyn. Range =20log(ΔB_{max}/ ΔB_{min})=9.8 dB	A_{totEff}=27170 μm^2.	[166]
7	$YBa_2Cu_3O_7$	100 (series)	1.2mm2	$S^{1/2}$≤4 pT/$Hz^{1/2}$ ΔV=2100 μV dV/dT=3.5 μV/nT, f=1kHz Estimations: ΔB_{max}=2100μV/3.5 μV/nT =600nT For Δf=1GHz, ΔB_{min}=126 nT Dyn. Range =20log(ΔB_{max}/ ΔB_{min})=13 dB	without magnetometer	[167]

N	Material	Number of Cells/Josephson Junctions	Area	Parameters,	Comments	Ref.
8	$YBa_2Cu_3O_7$	112 (series)	10*10mm	$S^{1/2}$=10 pT/$Hz^{1/2}$ ΔV=3500 μV dV/dT=600 V/T, f=1Hz-10GHz Estimations: ΔB_{max}=0.0035V/600 V/T =5800nT For Δf=1GHz, ΔB_{min}=316 nT Dyn. Range =20log(ΔB_{max}/ ΔB_{min})=25 dB		[159]
9	$YBa_2Cu_3O_7$	211 (series)	1.2mm2	$S^{1/2}$=0.4 pT/$Hz^{1/2}$ ΔV=2100 μV dV/dT=5000 V/T, f=1Hz-10GHz Estimations: ΔB_{max}=0.0021V/5000 V/T =420nT For Δf=1GHz, ΔB_{min}=12.6 nT Dyn. Range =20log(ΔB_{max}/ ΔB_{min})=30 dB	The on-chip superconducting pick-up loop enhances the SQIF magnetic field sensitivity of about 10 times with respect to the bare SQIF of the same geometry without the pick-up loop.	[182]
10	$YBa_2Cu_3O_7$	20 (series)		ΔV=12000 μV dV/dT=270 V/T, f=1-2GHz Power gain: 20 dB Estimations: ΔB_{max}=0.012V/270 V/T =44 μT		[168]
11	Nb	2*108 (series) each with 10 parallel		ΔV=22000 μV	Rn=1.6 Ohm.	[169]
12	Nb	128 (series)		ΔV=7000 μV dV/dT=50 V/T Estimations: ΔB_{max}=0.007V/50 V/T =140 μT	Bi-SQUIDs	[170]
13	Nb	2*84 (series) each with 10 parallel	3.3*3.3mm	dV/dT=750 V/T Linearity 70 dB	With flux transformer (concentrator)	[171]

Table 2-18. (Continued)

N	Material	Number of Cells/Josephson Junctions	Area	Parameters,	Comments	Ref.
14	Nb	1000 (series)	5*5mm	ΔV=140 mV Power gain: 25-30 dB Linearity 40 dB Output measured power level: -50-65 dBm	Bi-SQUIDs	[172]
15	Nb	2*560 each with 12 parallel	3.3*3.3mm	ΔV=100 mV dV/dT=6.5 mV/μT Linearity 70 dB Estimations: ΔB_{max}=100mV/6.5 mV/μT =15 μT		[173, 153, 152]
16	Nb	30(parallel) *80(series) =2400		Gain: 2-24 dB for f=300 kHz-1GHz	Single SQUIDs with areas of 72 μm^2 were used.	[175]
17	Nb	2*588 each with 12 parallel	4.2*4.5mm	ΔV=100 mV dV/dT=8000 mV/mT Linearity 70-100 dB Estimations: ΔB_{max}=100mV/8000 mV/mT =12.5 mT		[154]
18	$YBa_2Cu_3O_7$	484 (series)	10*10mm	$S^{1/2}_{v}$=40 nV/$Hz^{1/2}$, $S^{1/2}_{\Phi0}$=0.75 μΦ_0/$Hz^{1/2}$ ΔV=17 mV f=1kHz Estimations: $S^{1/2}$= $S^{1/2}_{\Phi0}$/A_{totEff} =1.87 pT/$Hz^{1/2}$ For Δf=1GHz, ΔB_{min}=59 nT For Δf=1GHz, ΔV_{min}=1.23 mV Dyn. Range =20log(ΔV/ ΔV_{min})=22.8 dB	With flux focuser, A_{totEff}=89*9 μm^2.	[176]

N	Material	Number of Cells/Josephson Junctions	Area	Parameters,	Comments	Ref.
19	$YBa_2Cu_3O_7$	770 (series)	10*10mm	$S^{1/2}_v$=8 nV/$Hz^{1/2}$, $S^{1/2}_{\Phi 0}$=0.25 $\mu\Phi_0/Hz^{1/2}$ ΔV=10.1 mV f=1kHz Estimations: $S^{1/2}$= $S^{1/2}_{\Phi 0}/A_{totEff}$ =0.34 pT/$Hz^{1/2}$ For Δf=1GHz, ΔB_{min}=10.7 nT For Δf=1GHz, ΔV_{min}=0.25 mV Dyn. Range =20log(ΔV/ ΔV_{min})=32.1 dB	With flux focuser, A_{totEff}=160*9 μm^2.	[176]
20	$YBa_2Cu_3O_7$	20,000 (total), 20 subarray, each subarray Np=50 (par)*Ns=20(ser)	10*10mm	$S^{1/2}$=? pT/$Hz^{1/2}$ ΔV=0.9 mV dV/dT=1530 V/T, f=30MHz Estimations: ΔB_{max}=0.0009/1530=0.7μT	A_{totEff}=20000*280 μm^2 =5.6 mm^2. I_c=12.7 μA, I_c(array)=172 μA, I_b=220 μA, R_n=5.34 Ohm, Z_0=42 Ω. β_l=1, without flux focuser	[177]
21	$YBa_2Cu_3O_7$	100,200(total), 6 subarray, each subarray Np=100 (par)*Ns=1670(se r)	10*10mm	$S^{1/2}$=10.9 pT/$Hz^{1/2}$ (estimated) ΔV=2.41 mV dV/dT=24800 V/T, f=?MHz Estimations: ΔB_{max}=0.0241/24800=0.97μT For Δf=1GHz, ΔB_{min}=345 nT Dyn. Range =20log(ΔB/ ΔB_{min})=8.9 dB	A_{totEff}=100200*30 μm^2 =3.06 mm^2. I_c=30 μA. β_l=0.4, without flux focuser	[148]
22	$YBa_2Cu_3O_7$	1,012,500 (total), 9 subarray, each subarray Np=500 (par)*Ns=225(ser)	10*10mm	$S^{1/2}$=? pT/$Hz^{1/2}$ ΔV=53.3 mV dV/dT=40100 V/T, f=?MHz Estimations: ΔB_{max}=0.0533/40100=1.3μT	without flux focuser	[148]

2.7.4. Short Summary of Experimental Data

The dynamic range of a sensor is the ratio between the largest and the smallest signals. For a SQIF without a feedback loop this ratio is given by $\Delta V_{max}/(S_V(f))^{1/2}$, where ΔV_{max} is the maximum voltage swing of the SQIF peak. As maximum voltage swing scales with number of SQUIDs (or Differential Quantum Cells), the dynamic range scales with the square root of SQUID number. For example, the dynamic range of a typical high-Tc SQIF with 100 loops is therefore 10 times (or 20 dB) larger than for a conventional single loop SQUID. Properly designed SQIF's can therefore be operated in open loop mode and could be directly impedance matched to a conventional preamplifier. Table 2-18 summarizes some design operational parameters of SQIF/bi-SQUID antennas.

2.7.5. Trends and Future Activities

Although SQUID/SQIF/bi-SQUID array antennas have achieved very impressive results and very promising applications for radars and communication systems will be surely found in the future, it is still too early to say that this technology reached a mature in order to be commercialized.

Present trends are mainly related to testing some design ideas on new SQUID/SQIF/bi-SQUID arrays and in some cases, such antennas are also tested as prototypes in radars or communication systems.

Based on the published papers in the opening literature sources, it seems that in the future main activities must be related to the systematic approach in several technological and engineering fields in order to make reasonable product for applications in radars and communication systems. Some fields could be summarized as follows:

- Though high-Tc materials fabrication is relatively advanced, still only few firms or scientific institutions worldwide have sufficient experience with production of SQUID/SQIF arrays.
- Systematic and detailed investigations on different types of SQUID, SQIF, bi-SQUID arrays, without or with flux focuser

(concentrator), or with external superconducting loop antenna is still missing, but will be greatly appreciated. It is still not clearly stated, which array types have advantages or disadvantages for applications in radars and communication systems.

- It would be very helpful, if high frequencies engineers with experience on design of sc components like filters or antennas, participate in the design of SQUID/SQIF/bi-SQIUD. Only in very limited publications, some high frequency design parameters have been mentioned. For proper RF design, many topics must be carefully considered, e.g., 50 Ω lines, impedance matching at interfaces, cross-talk (interference) between different lines, etc. What is also rarely mentioned is the overall design of receiver, i.e., SQUID/SQIF/bi-SQUID is a broadband antenna and at the same time it is also an amplifier, and, therefore, after that a filter (most probably superconducting due to low insertion losses) must be connected, which is further followed by LNA (cold or warm). For example, in many papers, this filter is not mentioned, but if very broadband signal after SQIF is directly applied to LNA without passing through filter, the total noise level (i.e., integrated over whole bandwidth) could be extremely high, even higher than one with conventional warm antenna and electronics. For that reason, it is necessary to make careful RF design of the receiver, and most probably, the chip with SQUID/SQIF/bi-SQUID antenna will also incorporate sc filter.

For the successful and very promising applications of SQUID/SQID/bi-SQUD antenna for high frequency applications, it is still necessary to solve several engineering and technological tasks particular for high-Tc superconductors, before commercialized products will be available on the market.

2.8. Filters

First, it is worth to note that sc filters are still the "workhorse" of any receiving or transmitting radars or communication systems, which are designed for operation at low temperatures. If decision to apply cryogenic temperatures is taken, it is practically unavoidable to use sc filter. In some cases, only sc filter is placed at low temperatures, though for the most practical applications, at least sc filter and LNA are cooled.

Main application advantages of superconducting filters could be summarized as follows:

- Small size,
- Light weight,
- Very low in-band insertion loss,
- Very high off-band rejection,
- Flat in-band group delay
- Steep skirts.

It is worth to pay attention to one important difference in applications of superconductor filters [207]. For classical receiving systems, a LNA is connected after antenna, which has disadvantage that this amplifier is broadband, and, therefore, will receive and process all signals, including noise and side lobes of antenna. In some cases, which are not rare, this can lead to the severe operational challenges, if large out-of-band signals are present, which often leads to amplifier saturation or to intermodulation products in the band of interest. If a narrow filter before LNA is placed, then such interference could be significantly reduced, and only expected in-band signals are sent to LNA, and, therefore, the selectivity of system is improved. However, this approach is rarely done, because of additional losses in filter, which can significantly reduce receiver sensitivity. For superconducting filters, which have very low losses, in the order of 0.5-0.01 dB, it is reasonable to place them before LNA, which is done for many (practically near all) applications cases [207]. So, applications of the superconducting filters bring additional advantages for overall system performance.

And the last but not the least, looking at enormous number of different high-Tc filters, which have been successfully built, the author is convinced that “nothing is impossible.”

2.8.1. General

Development of high-Tc superconducting filters started practically immediately after discovery of high temperature superconducting materials in the second half of 1980s. Geometries of sc filters are practically the same as ones made from conventional materials, mainly aluminum or copper, see Figure 2-42 below for just very few examples.

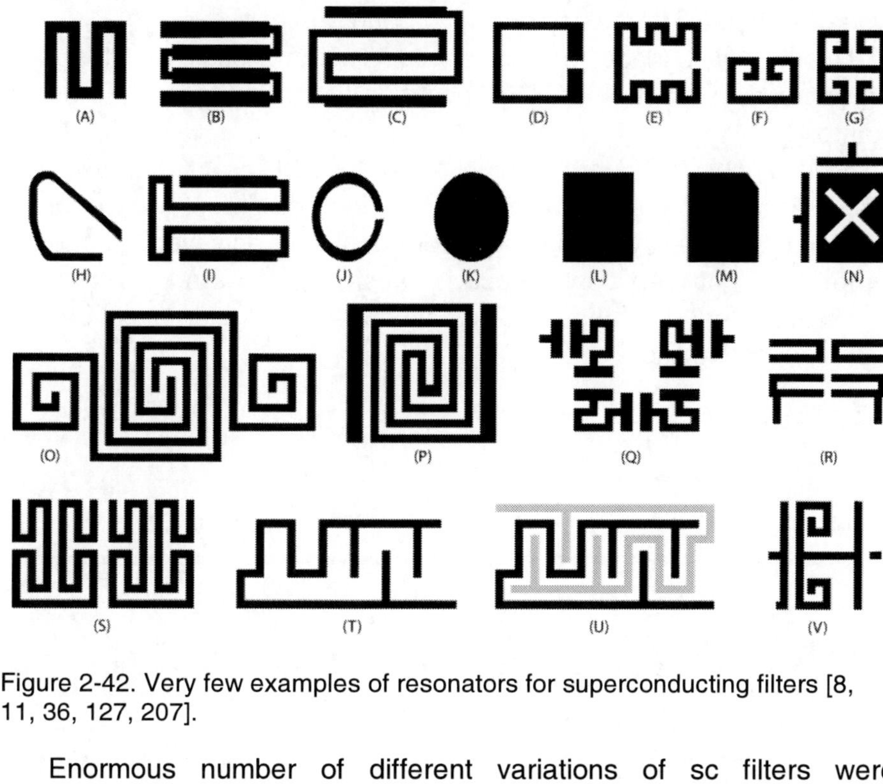

Figure 2-42. Very few examples of resonators for superconducting filters [8, 11, 36, 127, 207].

Enormous number of different variations of sc filters were developed over the last 30 years, and it is simple impossible to give a detailed discussion of all sc filters. For that reason, only few examples are mentioned and for more details, some books or review articles should be considered [6, 207, 228, 232]. Due to a high application potential, more attention is given to the tunable filters. It is also worth to note, that over the last 30 years, the development activities are shifted from European and North America region to Asia-Passific one, and many interesting versions are done in China and Japan.

2.8.2. Non-Tunable Filters

A short summary of insertion losses for different high-Tc filters, which have been manufactured over the last 5-10 years, could be found in [6]. For filters with designed number of poles 4 or more, the typical insertion losses were below 0.5 dB.

It is also worth to mention that filters with very small Fractional Band Width (FBW) below 0.1% (typically in the range 0.01 till 0.06%) were designed for frequencies up to C-band [6].

Due to low surface resistance, high-Tc filters could be designed with large number of poles, e.g., above 20 [6], see Figure 2-46 as an example.

Skirt slopes could be as high as 120 dB/MHz or 30dB/kHz and wideband rejection up to 100 dB [6]. The in-band group delay of order 20-50 ns over nearly whole bandwidth could be achieved, through a compromise between it and selectivity must be considered.

Figure 2-43 till Figure 2-46 show some examples of the superconducting filters.

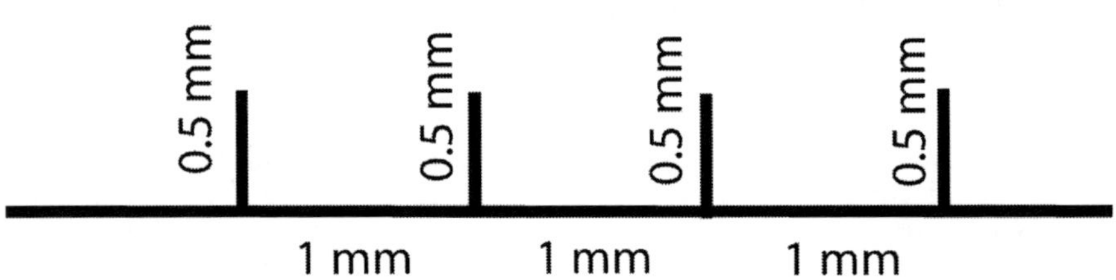

Figure 2-43. The simplest version of superconducting filters working in the range 1-10 GHz [259].

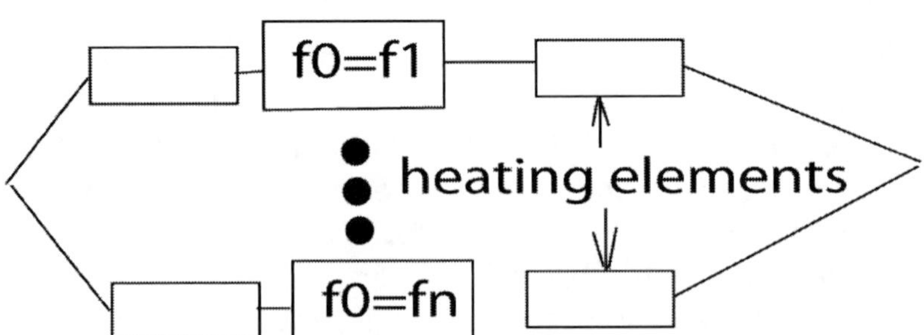

Figure 2-44. Switched Filter Banks: simplified presentation [229].

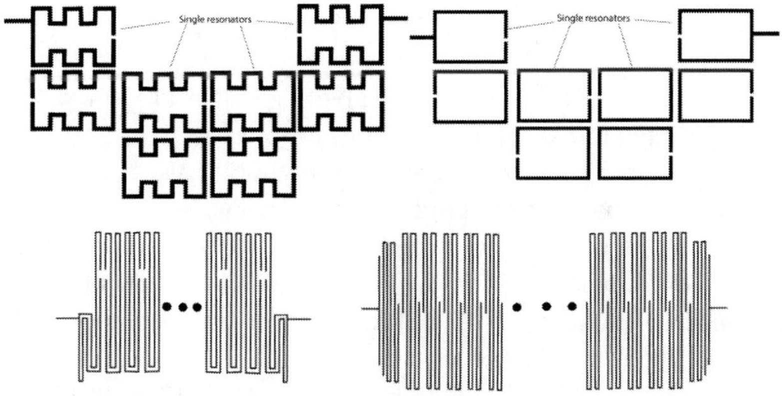

Figure 2-45. Layout of the HTS microstrip filter on an LAO substrate [8, 207].

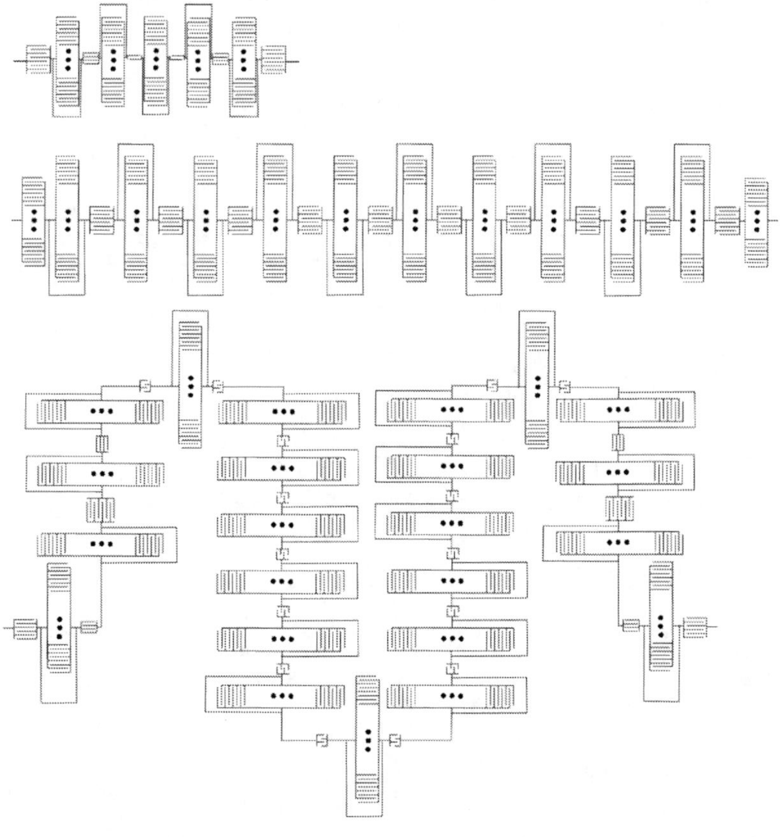

Figure 2-46. 5-, 10- and 23- pole filters on 2-inch substrate [230].

2.8.3. Duplexer and Multiband Filters

Due to advantageous properties of high-Tc materials, the duplexers and multiband filters can be designed to be of small size and mass as well as to have low insertion losses and very high selectivity [233-243]. For example, a manifold-coupled duplexer consisted of a T-junction, band-stop filter, and two band-pass filters with more than eight poles [6, 236]. Insertion losses were below 0.2 dB, skirt – 21 dB/MHz, out-of-band-rejection – more than 75 dB, insulation – more than 80 dB and voltage standing ratio – better that 1.4:1.

Table 2-19 shows some example of sc diplexers [228, 233].

Table 2-19. Superconducting diplexers applied for transceivers for communication systems

N	Center Freq., Channel 1, MHz	Center Freq., Channel 2, MHz	Insertion loss (dB)	Isolation (dB)	Return Loss (dB)	Size (mm)	Reference
1	216	220	0.16	>76	17	41.6*28.0	[235]
2	1800	2450	0.94	>55	14	22.0*34.0	[236]
3	2820	2855	0.35	>47	15	32.6*18.0	[237]
4	483	500	0.13	>60	18.9	22.9*21.7	[233]
5	1800	2400	0.05	>40	18.5	11.1*8.43	[239]
6	1800	2400	0.41	>39			[240]
7	1900	2100	0.19	>53		20*14	[242]

The next important class of sc filters is multiplexers, i.e., duplexer, triplexers and quadruplexers, see Table 2-20 below.

Table 2-20. sc du-/tri- and quadruplexers

N	Center Frequencies	Frequency	Insertion loss (dB)	Isolation (dB)	Return Loss (dB)	Size (mm)	Reference
1	3912 MHz 4078 MHz	1.5% 1.5%	0.1 0.1	40 dB		30*10	[260]
2	1200 MHz 1250 MHz	1.6% 0.8%	0.2 0.2	>80	15.6 15.6	42.0*26.0	[234]
3	1778 MHz 1842 MHz	0.8% 4%	0.3-1 0.3-1	>35	30	13*38	[244]

N	Center Frequencies	Frequency	Insertion loss (dB)	Isolation (dB)	Return Loss (dB)	Size (mm)	Reference
4	1200 MHz 1260 MHz	2.3% 2.2%	<0.2 <0.2	>90	>14 >14	42.0*26.0	[261]
5	2800 MHz 2810 MHz 2820 MHz 2830 MHz	0.2% 0.2% 0.2% 0.2%	0.32- 0.45	>55	>19 >19 >19 >19		[262]
6	9.0 GHz 9.12 GHz 9.2 GHz		<0.7				[263]

Two-three and four-bands filters are also developed [6, 232]. In order to obtain a multi-band filters, several serially connected dual filters could be used, though more advanced designs use multi-mode resonator filters, see Table 2-21 below for just few examples [232].

Table 2-21. Few examples of multi-band filters

	Short description	Frequency	Fractional Bandwidth	Insertion loss (dB)	Return Loss (dB)	Size (mm)	Refer.
1	Dual-band filter with Split Ring Resonators	2.45 GHz 3.5 GHz	18.29% 14.57%	0.37 0.35	>12 >13	12.1*8.4	[232, 265]
2	Tri-band filter with stepped impedance Split Ring Resonators	1.57 GHz 3.5 GHz 5.8 GHz	14% 4.86% 5.17%	0.15 0.54 0.36	>29 >11 >12	11.3*13.35	[232, 264]
3	Dual-band filter with multi-mode Split Ring Resonators	1.60 GHz 2.42 GHz	9% 7.4%	0.1 0.09	>16.1 >17.7	10.6*13.75	[232]
4	Tri-band filter with Stub-loaded Resonators	2.45 GHz 3.5 GHz 5.2 GHz	9.7% 5.1% 1.9%	0.16 0.55 0.22	>16.7 >10.2 >17.5	8.3*8.6	[232]
5	Tri-band filter with Stub-loaded Resonators	1.57 GHz 2.6 GHz 3.5 GHz	3.4% 1.3% 0.5%	0.58 0.64 0.61	>10.7 >11 >11.6	9.2*7.6	[232, 266]

Table 2-21. (Continued)

	Short description	Frequency	Fractional Bandwidth	Insertion loss (dB)	Return Loss (dB)	Size (mm)	Refer.
6	Quad-band filter with Stub-loaded Resonators	1.57 GHz 2.1 GHz 4.6 GHz 5.1 GHz	15% 7.1% 3.2% 2.9%	0.13 0.07 0.135 0.2	>32.3 >19.4 >23.7 >16	11.1*14.95	[232]
7	Quad-band filter with Square Ring Loaded Resonator, second order	2.43 GHz 3.53 GHz 5.18 GHz 5.78 GHz	4.96% 5.07% 2.32% 3.63%	0.12 0.12 0.23 0.25	>20 >20 >15 >16	4.8*18	[232]
8	Dual Band	1.8 GHz 2.4 GHz	5.9% 4%	0.1 0.13	>12 >13		[245]
9	Dual band bandpass filters with bandstop notches. Notch filters has 0.1 till 1% bandwidth	0.37 GHz 0.41 GHz	7.5% 10.2%	0.1 0.1	>18 >16	32*30	[246]
10	Dual band filter	3.0 GHz 3.06 GHz	1% 0.98%				[247]
11	Dual band filter	1.93 GHz 1.97 GHz	1.0% 1.0%	0.2 0.2	>15 >15	>20*20	[249]
12	Dual band filter	3.45 GHz 4.9 GHz	8.7% 4.1%	0.2 0.3	>15 >14	19.3*8.4	[250]
13	Dual band filter	1.11 GH 1.23 GHz	1.4% 1.3%	0.22 0.13	>15 >15		[251]
14	Dual band filter	2.55 GHz 3.53 GHz	6.9% 3.8%	0.3 0.33	>16 >16	18.14*9.36	[252]
15	Dual band filter	1.8 GHz 2.4 GHz	4.4% 1.25%				[253]
16	Dual band filter	2.15 GHz 3.50 GHz	2% 2%				[254]
17	Tri- band filter	3.5 GHz 4.25 GHz 5.0 GHz	2% 2%				[255]
18	Dual band filter	2.4 GHz 5.8 GHz		0.12 0.18			[256]

	Short description	Frequency	Fractional Bandwidth	Insertion loss (dB)	Return Loss (dB)	Size (mm)	Refer.
19	Dual band filter	1.9 GHz 2.6 GHz				17.6*14	[257]
20	Tri- band filter	0.877 GHz 0.882 GHz 0.886 GHz	0.26% 0.26% 0.26%		>25 >25 >25	40*20	[248]
21	Tri-band filter with embedded Stub-loaded Resonators	1.8 GHz 2.4 GHz 3.5 GHz	4% 4% 4%	0.07 0.08 0.11			[267]
22	Tri-band filter with a pair of asymmetric shunted-line stepped impedance resonators	1.575 GHz 2.4 GHz 3.5 GHz		0.1 0.1 0.1			[268]
23	Tri-band filter with asymmetric trisection stepped-impedance Resonators	1.5 GHz 3.5 GHz 5.8 GHz		0.16 0.16 0.16			[269]

2.8.4. Tunable Filters

Some first versions of superconductor tunable filters were designed as a parallel combination of single filter lines, and selection of required line was done by “switching” of other into normal conducting state with electrical heaters, see Figure 2-44 as an example.

Presently, switching between single lines is performed “electronically” by using either ferroelectric, semiconductor or radio-frequency micro-electro-mechanical system (RF-MEMS) tuning technologies. The RF-MEMS technology offers low losses, which are typically in the range 0.5-4 dB, and possibility to digitization by using MEMS switches. Figure 2-47 shows the 4-pole tunable filter with semiconductor varactors [6].

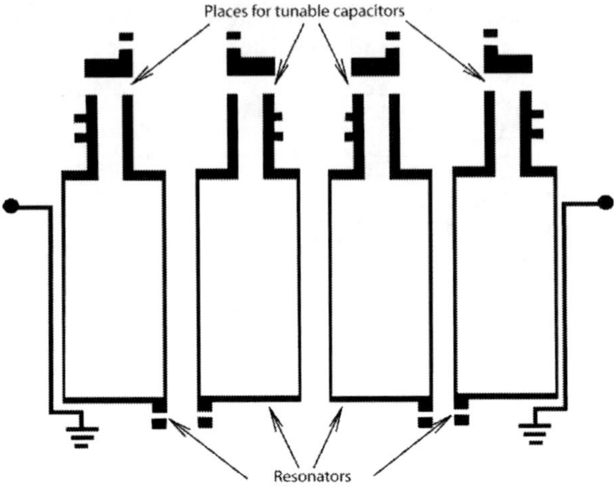

Figure 2-47. 4-pole tunable filter (with semiconductor varactors. The insertion loss was in the range of 0.8-3.8 dB, and return loss was better than 10 dB.

Table 2-22 shows a short summary of superconducting tunable filters, which were developed over the last 20 years. The insertion losses are somehow higher than the ones of non-tunable filters, which is related to the fact, that *Q* of resonators is reduced due to applications of the tunable elements.

Table 2-22. Short summary of superconducting tunable filters

N	Type	Frequency and Tunable Ranges	S21	S11	BW	Year & Ref.
1	Ferroelectric (BaTiO3),	3.5 GHz,	0.05 dB			2003, [209]
2	MEMS, two-pole	0.58-0.755 GHz, Δf =0.175 GHz (25%)	0.7-1 dB		around 1 MHz (0.14%)	2005, [210]
3	Ferroelectric (BaSrTiO3),	11.74-11.93 GHz, Δf =0.19 GHz (1.6%)	0.35-1.6 dB		0.47 GHz (4%)	2006, [211]
4	MEMS, to switch between two filters with 2.1 and 2.6 GHz and BW=200 MHz	2.1 or 2.6 GHz filter	2-4 dB		0.2 GHz (8.5%)	2007, [212]

N	Type	Frequency and Tunable Ranges	S21	S11	BW	Year & Ref.
5	Ferroelectric (BaSrTiO3), three pole.	3.50-3.68 GHz, Δf=0.18 GHz (5.1%)	1.4 dB		0.208 GHz (5.8%)	2008, [213]
6	Mechanically tuned.	167.2-166.5 MHz, Δf=0.7 MHz (16%)	0.4 dB		0.54 MHz (0.3%)	2009, [214]
7	Ferroelectric (BaTiO3), two pole	6.5 GHz, Δf=0.228 GHz (3.5%)	2-3.5		100 MHz (1.5%)	2013, [215]
8	Mechanically switchable,	198.2-199.2 MHz, Δf=1 MHz (16%)	0.69 dB (max)		400 kHz (0.2%)	2013, [216]
9	Notch filter, GaAs varactor,	286-485 MHz, Δf=199 MHz (3.5%)	0.2 dB (out-of-band)		6.9 to 7.9 MHz (2.55-2.44%)	2014, [217]
10	GaAs varactor, 2 or 4 poles,	430-720 MHz, Δf=290 MHz (50%)	0.8-3.8 dB	10 dB	17 MHz (3%)	2014, [218]
11	Dual mode operation, GaAs varactor, 4 poles,	784.92 – 918.23 MHz, Δf=133 MHz (16%)	0.69 dB (max)	9.3 dB	170 MHz (20%)	2014, [219]
12	GaAs Varactor, 2-poles	924.5 – 1030.8 MHz, Δf=106.3 MHz (11%)	0.16-1.4 dB		23.8 MHz (2.4%)	2014, [220]
13	Varactor, 4 poles.	983.85 – 1103.6 MHz, Δf=119.7 MHz (11.5%)	0.46-2.11 dB		29.3-60 MHz (2.8-5.7%)	2014, [271]
14	Varactor, 2 poles.	274.1 – 317.7 MHz, Δf=43.6 MHz (14.4%)	0.08-0.7 dB		5.14 MHz (1.74%)	2017, [221]
15	Varactor, 5 poles.	235 – 250 MHz, Δf=15 MHz (6.2%)	2.5-9.4 dB		0.7 MHz (0.29%)	2019, [222]

2.8.5. High Power Filters

Development of high-power superconductor filters was mainly related to an application of either film resonators or bulk ones, which are manufactured from superconducting materials and inserted into some dielectric material. In the latter design, the signal is coupled to sc resonator, transferred between resonators by weak link over dielectric material and tuning is performed by mechanical adjustment/movement of superconducting materials.

One of the first successful realization of microstip resonators for 1.9 and 3.6 GHz with corresponding Q_0 values of 13000 and 6000 respectively, is presented in [223]. The stored energy within the resonators was around 12 kW, and for 5-pole filters, the transferred energy is around 50 W. Application of 3- and 4-channels multiplexers and filters for the satellite communication with power handling of 40W per filter was also demonstrated [224].

Quite remarkable results of 115 and 74 W were achieved on 2-poles and 3-poles octagonal resonators with center frequency of 2.88 GHz [231]. It is worth to note that larger power levels were not applied due to power limitation of available RF generator.

Further developments of planar filters could be found in [225], where a four-pole filter with 40 MHz bandwidth for frequency range 3.4-4.2 GHz and power handling of 60 W was designed.

A successful development of non-strip line filter could be found in [226]. Such two-pole filter with Q_0=85000 could handle 35 W power at 6.23 GHz and bandwidth of 1%. It is worth to note that larger power was not applied due to power limitation of the available RF generator.

Disk- and ring-type bulk resonators are presently also under active development [241].

Table 2-23 summarizes some filter parameters of ring/disk type resonator type.

Table 2-23. Filter dimensions and some operating parameters [241]

	Disk Filter	Ring Filter
Thickness of resonator	0.5	0.5
Outer diameters of 1st and 3rd resonators	8.2	7
Outer diameter of 2nd resonator	9.48	8.1
1st and 3rd substrate size	19.8*10.2*6.5	17*9*6.5
2nd substrate size	19.8*11.48*6.5	17*10.1*6.5
Cavity size	28*43.8*19.6	25.2*38.3*19.6
Power handling capability	41.7 W	26.6 W

It is worth particular to underline that for micro-strip filters with resonators and moderate power transfer up to 100W, very high powers of order several tens of kilowatts (up to 50 kW [227]) is stored within resonators. This is related to the fact that due to low losses, very high Q_0 values $2 \div 60 \cdot 10^4$ are achieved.

2.8.6. Trends and Future Activities

At the present, it is possible to say that development of sc filters reached a mature state. Worldwide, many groups achieved remarkable results, and superior of sc filter over conventional ones was shown for nearly all sc filters built over the last 20 years.

However, it is worth to particularly stress, that for the successful design of sc filters, it is necessary to have some knowledge in high frequency properties of superconducting materials, as well as in production technology.

Presently, it is possible to build any band-pass, band-stop, notch filters or its combination.

Somehow, the tunable filters are less developed, but they are actively investigated. It is possible, that with the development of tunable filters, even more flexible operation of radars and communication systems will be achieved.

2.9. Power Limiters

Development of sc power limiters for receivers was also done [272-280]. Two main design principles are applied, i.e.:

- Making the limiter completely from sc material. After transition to the normal conducting state, the impedance of superconductor is significantly increased, which limits the transmission [272-276]. This concept is similar to the current limiters, which are applied for the high current limitations at superconducting devices, e.g., at coils or transformers.

- Making the limiter mainly from copper and only small part is designed from superconductor [277-280]. In this case, the limiter has filter structure. If sc part has high resistance, the quality factor of resonators is significantly reduced, which leads to decoupling of resonators and incident high power signal is effectively reflected back and only small power is adsorbed inside the limiter. If sc part has low resistance, the resonators are effectively coupled and quite significant power could be further transferred. Operation ranges for transmission could be around 10 mW, while for limitation one - up to 40 W.

For the practical applications, it is possible to choose either approach.

2.10. Other Developments

In this chapter, overall work done over the last 10-15 years on analog/digital SQUID-based amplifiers or mixers is overviewed. Some prototypes built and published in the opening literature sources are also mentioned. It could be possible that some of developments could find wider applications in the future.

2.10.1. Analogue SQUID

In the recent years, few groups presented designs of analog SQUID, which could be operated in GHz range [183-188]. The progress of analog SQUID development was mainly related to the advantages of SQUID fabrication process. The SQUID is operated in open loop mode, i.e., without feed-back loop.

Some versions of analogue SQUIDs working in the high frequency ranges, up to 10-15 GHz (they are also named as Superconducting Quantum Amplifiers (SQA)) are listed in Table 2-24. In this case, no feed-back loop is applied and coupling of RF-signal to SQUID and from SQUID to further analog electronics was realized with single-layer slot and coplanar lines [183-188]. In some cases, instead a coil, a micro strip line is used as a resonator [285].

It is worth to note that for the amplification of weak signals, which are typically in the range 100-400 μV, a relative complicated circuit with output filter is applied [183]. The relative low gain and dynamic range are still limiting their applications for the radars or communication systems, so semiconductor amplifiers, which have comparable characteristics, are typically applied.

Table 2-24. Some operational parameters of analogue SQUIDs

N	Gain, dB	Noise Temp., K	Bandwidth, 3dB (MHz)	Frequency (GHz)	Dynamic range (P_{out}/P_{in})	Equivalent input L_i and R_i (excluding input matching network)	Ref.
1	9.8	2.5	300	3.95	4.5	Li=3 nF [188]	[183]
2	11.5	1.0	500	4.0	7.5		[185]
3	11-12	5-10	500-1000	11	10	Li=50pF, Ri=0Ω [184] Li=50-250pF, Ri=0Ω [187]	[184]
4	4.5		70 (high-Tc filter was used before SQUID)	1.75			[186]
5	11	4.0	300	3.65			[282, 283]
6	17.5	4.0	250	3.65			[283, 284]
7	12 11 10 9 6	1.3 1.5 1.8	500 550 300 320 850	2.2 3.2 4 5 7.4			[285]
8	17	0.55	150	3.8			[286]

One additional limitation on application of analog SQUID is so called input saturation temperature ($T_{satur}=P_{inpSat}/k$), which described maximal input power at which SQUID output is saturated. For low-Tc Nb-based superconductor it is around 55 K□GHz, while for high-Tc YBCO-based superconductor – 500-1000 K□GHz. For example, for the bandwidth of 1 GHz, the equivalent inlet temperatures are 55K and 500-1000K for low- and high-Tc superconductor SQUIDs, respectively.

It is also important to keep in mind two important design and operational parameters, i.e.:

- To maintain a reasonable degree of Josephson coupling, it is necessary to have the operating current above the one defined by thermal noise, i.e., I_{th}≈3.3 μA for 77K, and 0.17 μA for 4.2K, otherwise the Josephson coupling energy will be destroyed. For Nb-technology, the $I_{operation}$ is typically chosen as 125 μA.
- The intrinsic white noise of the SQUID loop should be sufficiently low not to allow destruction of magnetic flux, which is stored inside. So, the loop inductance must be below L_{th} ≈321 pH for 77K, and 5.9 nH for 4.2K.

2.10.2. Digital SQUID and Digital DROS

The developments of digital SQUIDs and DROS working in high frequency ranges, up to 3 GHz are mentioned in this chapter [191-201]. In this case, the feed-back loop is applied (i.e., closed-loop operation), and coupling of the RF-signal to SQUID/DROS is done by the signal coil, which is further connected to the antenna pick-up coil. The output of the digital SQUID/DROS is a digitized signal. Figure 2-48 below shows the two versions of the digital SQUID.

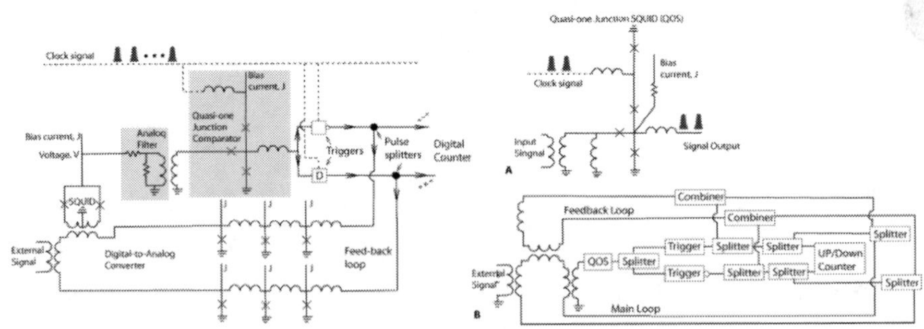

Figure 2-48. Digital SQUID concept with SQUID and Quasi-One-Junction Comparator (QOC) (left) [201], and only QOS comparator with single flux quantum (SFQ) readout terminals and two types of SFQ feed-back loops (right) [200].

It is worth to note that digital SQUIDs are very similar to the oversampling delta-type (Δ-type) ADC based on Rapid Single Flux Quantum (RSFQ) logics. Sub-Φ_0 resolution with corresponding feed-back loop has already been used for the oversampling ADC to reduce quantization noise. The digital SQUID, presented in Figure 2-49 (right), also included the decimation filter for fast data transfer to room temperature electronics as well as synchronization channel. Though the pulse generation frequency was 66 GHz, the clock frequency was only 5 GHz and it is expected that operation at higher frequencies will be possible in the future. Due to realization of two feed-back loops, it was possible to keep the currents in these loops at constant value, and practically unlimited operational range on magnetic field was also realized.

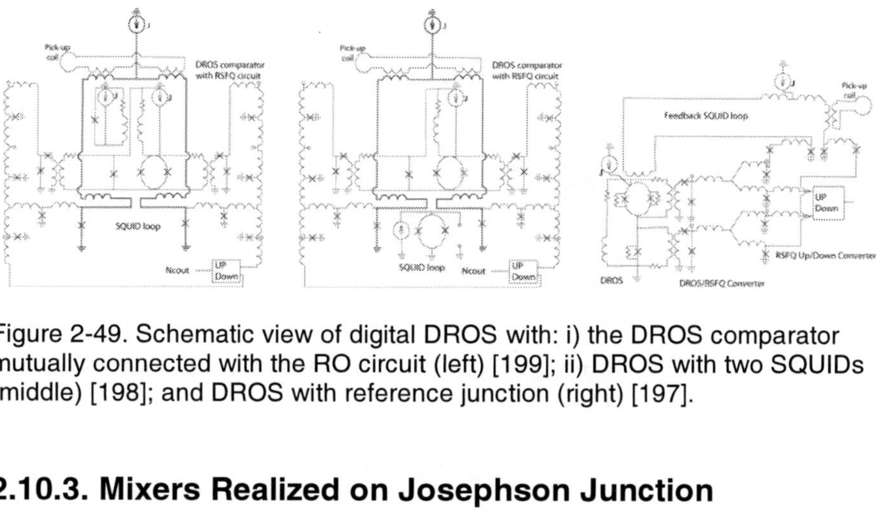

Figure 2-49. Schematic view of digital DROS with: i) the DROS comparator mutually connected with the RO circuit (left) [199]; ii) DROS with two SQUIDs (middle) [198]; and DROS with reference junction (right) [197].

2.10.3. Mixers Realized on Josephson Junction

In recent years, several groups developed down-converters based on application of Josephson Junction and operating at G- and THz frequency ranges, for example:

- superconducting detector-based THz imager working at 200-600 GHz [142, 145, 146, 288, 289],
- frequency-tunable gigahertz heterodyne oscillator and mixer for frequencies up to 5 GHz [142, 143, 144, 288, 289],

- Mixers for digital spectrometers working up to 1 THz [202, 203, 288, 289],
- Mixers for receivers used in Radio-Astronomy for frequencies up to 1 THz [189, 190, 204-206].

Though several systems have been developed, this application field is small and very specialized. From the radars and communication systems, working in the frequency range 500 GHz till 1 THz or above they are quite interesting. A promising future application could be "combination" of mixer, working on Josephson Junction, and superconducting electronics based on RSFQ logics [190, 202]. For that, a prototype system must be designed and thoroughly investigated.

2.10.4. Trends and Future Activities

At the time being, either some prototypes or very specialized products have been developed. As the marked is still limited, it is not expected that many new systems will be constructed in the nearest future.

For the future activities, it will be quite interesting to design a radar or communication system working in the frequency range of 500 GHz till 1 THz or above, which are based on mixer working on Josephson Junction and superconducting electronics based on RSFQ logics. For such system, the signal will be directly digitized after mixer.

References

[1] Klauda M., et al., (2000). Superconductors and Cryogenics for Future Communication Systems, *IEEE Trans. Microwave and Techniques*, 48:1227-1239.

[2] Nisenoff M., (1996). The high-temperature superconductivity space experiment (HTSSE): present status and future challenges, *Supercond. Sci. Technnol.* 9:A174-177.

[3] Superconducting Technologi Inc. (STI), https://www.suptech.com/.

[4] Willemnsen B., (2009). Practical cryogenic receiver front ends for commercial wireless application, *International Microwave Symposium Digest*, *MTT-S*:1457-1460.

[5] HYPRES Inc., https://www.hypres.com/products/cryogenic-analog-rf-module/.

[6] Li Chunguang, et al., (2017). Progress on applications of high temperature superconducting microwave filters, *Supercond. Sci. Technnol.*, 30:073001.

[7] Knack A., et al., (2007). Superconducting Technology for Wireless Communication: Development of a CDMA Base Station Cryogenic Front End Receiver, *Proc. Of Asia Pacific Microwave Conference*, p187-190.

[8] Andrian Knack, (2006). *Design and Implementation of HTS technology for cellular base stations*, PhD Thesis, Massey University, New Zeeland and James Cook University, Australia.

[9] Jin J. X., et al., (2014). Enabling High-Temperature Superconducting Technologies Toward Practical Applications, *IEEE Trans. Appl. Supercond.* 24:5400712.

[10] Zang X., et al., (2010). Development of a HTS transceiver sub-system for 3G mobile communication TD-SCDMA base Station, *Supercond. Sci. Technnol.* 23:025007.

[11] Li C., et al., (2013). Recent Progress of HTS microwave applications in satellite receiver, meteorological radar, mobile communication and radio astronomy, *J Supercond Nov Magn.*, 26:1843-1848.

[12] Kayano H., et al., (2017). RF noisy environment measurement and throughput measurement by superconducting cryogenic receiver, *Asia Pacific Mircrowave Conf.*, p626.

[13] Kayano H., et al., (2011). Narrow-band superconducting hybrid filter for 100 kW transmitter of X-band radar applications, *International Microwave Symposium Digest, MTT-S*, p1-4.

[14] Kayano H., et al., (2011). Superconducting transmitting-receiving bandpass filters for 5 GHz band weather radar applications, *Proc. China-Japan Joint Microwave Conf.*, p1-4.

[15] Kawaguchi T., et al., (2011). Narrowband superconducting receiving filter for 9 GHz-band weather radar, *Proc. 41st European Microwave Conf.*, p744-747.

[16] Lancaster M., et al., (1996). Miniature Superconducting Filters, *IEEE Trans. Microwave Theory and Techniques*, 44:1339.

[17] Zhang G., et al., (2005). Superconducting microstrip bandstop filter for an L-band telescope receiver, *Proc. 35 Europ. Microwave Conference*, 8977994.

[18] Zhang G., et al., (2005). Superconducting spiral filters with quasi-elliptic characteristic for radio astronomy, *IEEE Trans. Micrwave Theory and Technique*, 53:947.

[19] Zhow J., et al., (2005). HTS narrow band filters at UHF band for radio astronomy, *IEEE Trans. Appl. Supercond.* 15:1004.

[20] Zang G., et al., (2006). High-temperature superconducting bandpass filter with microstrip quarter-wavelength spiral resonators, *IEEE Trans. Microwave Theory and Techniques*, 54:559.

[21] Niu D-C., et al., (2005). An X-Band front-end module using HTS technique for a commercial dual mode radar, *IEEE Trans. Appl. Supercond.* 15:1008.

[22] Seaux J. F., et al., (2004). Interest of the superconductivity at 30 GHz - Application to the HTS preselect receive filters for satellite communications, *International Microwave Symposium Digest*, MTT-S, p1121-1124.

[23] Shivare J. P. et al., (2002). High-temperature superconducting microwave filters for communication payloads of satellites, *Supercond. Sci. Technnol.* 15:983-985.

[24] Mansour R., et al., (2000). A 60-channel superconducting input multiplexer integrated with pulse-tube cryocooler, *IEEE Trans. Microwave Theory and Techniques*, 48:1171.

[25] Wu J., et al., (2013). Developments of microwave HTS-based devices and subsystems for applications in civilian satellites, *MSMW*, 8:46-51.

[26] Polturak E., et al., (1999). Design and Performance of a space based high temperature superconductivity experiment, *App. Phys. Lett.* 75:1293-1295.

[27] Kayano H., et al., (2014). 8 Elements phased array antenna using superconducting receiving module, *The 8th European Conference on Antennas and Propagation*, p1852-1853.

[28] Kayano H., et al., (2013). 3 GHz band HTS Multichannel Receiving Unit with 8 modules, *IEEE Radio and Wireless Symp. Dig*. p64-66.

[29] Kayano H., et al., (2014). Low-profile high-sensitivity sub-array module with HTS filters for an active phased array antenna, *IEEE Radar Conference (RadarCon)*, 14526852.

[30] Kumamoto T., et al., (2013). High-Sensitivity Receiving Sub-array Module with HTS filters for an active phased array antenna, *IEEE International Symposium on Phased Array System and Technology*, p341-345.

[31] Iijima K., et al., (2015). Superconducting sub-array module as T-R module for X-band active phased array antenna, *2015 IEEE Radar Conference (RadarCon)*, 15240070.

[32] Iijima K., et al., (2018). Compact superconducting sub-array module for X-band phased array antenna, *IEEE Radar Conference (RadarCon)*, 17842586.

[33] Kanaya H., et al., (2003). Design and performance of high Tc superconducting coplanar waveguide matching circuit for RF-CMOS LNA, *IEICE Trans. Electron.*, E86-C:p37-42.

[34] Kanaya H., et al., (2003). Design of HTS coplanar waveguide matching circuit for low noise CMOS-HTS receiver, *IEEE Trans. Appl. Supercond.* 13:1031.

[35] Kerans A., et al., (2014). High Temperature Superconducting Filters to Enable High Power Public Safety Mobile Broadband, Proceedings of Asia-Pacific Micrwave Conference, p402.

[36] Mazierska J., et al., (2004). "High temperature superconducting filters for wireless communication," a book chapter in *Novel Technologies for Microwave & Millimeter-Wave Applications,* Kluwer Academic Plenum Publishers, pp 123-152.

[37] Mazierska J., (2004). Superconducting Cryogenic Front End, *Prod. of MIKON, 15th Int. Conference on Microwaves, Radar and Wireless Communications*, p351-354.

[38] Mukhanov O., et al., (1989). RSFQ logic arithmetics, *IEEE Trans. on Magnetics*, 25:857.

[39] Mukhanov O., et al., (1987). Ultimate performance of the RSFQ logic circuits, *IEEE Trans. on Magnetics*, 23:759.

[40] Likharev K., et al., (1991). RSFQ Logic/Memory Family: A new Josephson-Junction Technology for sub-terahertz clock frequency digital systems, *IEEE Trans. Appl. Supercond.* 1:3-28.

[41] Likharev K., (1984). *Dynamics of Josephson Junctions and Circuits*, Gordon and Breach Science Publishers.

[42] Polonsky S., et al., (1993). New RSFQ circuits, *IEEE Trans. Appl. Supercond.* 3:2566-2577.

[43] Bunyk P., et al., (2001). RSFQ technology: physics and devices, *Int. J. of High Speed Electronics and Systems*, 11:257-305.

[44] Brock D., (2001). RSFQ technology: circuits and systems, *J. High Speed Electron Syst.* 11:307-362.

[45] Mukhanov O., (2015). *Digital Processing, Superconductor Digital Electronics*, in *Applied Superconctivity: Handbook on Devices and Applications* (ed. P. Seidel), Wiley-VCH Verlag GmbH & Co. KHaA, Weinheim, Germany, p1336, ISBN: 978-3-527-41209-9.

[46] Hayakawa H., et al., (2004). Superconducting digital electronics, *Proc. of IEEE*, 92:1534-1548.

[47] Sarwana S., et al., (2011). Multi Band Digital RF Receiver, *IEEE Trans. Appl. Supercond.* 19:667.

[48] Filippov T., et al., (2011). Multi-bit mixer for Digital-RF Receivers, *IEEE Trans. Appl. Supercond.* 21:818.

[49] Gupta G., et al., (2011). Modular, Multi-function Digital-RF Receiver System, *IEEE Trans. Appl. Supercond.* 21:883-890.

[50] Vernik I., et al., (2009). Progress in the Development of Cryocooled Digital Channelizing RF Receivers, *IEEE Trans. Appl. Supercond.* 19:1016-1021.

[51] Mukhanov O., et al., (2009). Digital RF Receiver Systems, *12th International Superconductive electronics conference*.

[52] Kirichenko D., et al., (2009). Microwave receivers with direct digitization, *IEEE MTT-S International Microwave Symposium Digest*, p1449-1452.

[53] Mukhanov O., et al., (2008). Superconductor Digital-RF Receiver Systems, *IEICE Trans. Electron.* E91-C:306-315.

[54] Vernik I., et al., (2007). Cryocooled Wideband Digital Channelizing RF Receiver Based on Low-pass ADC, *Super. Sci. Tech,* 20:323-327.

[55] Gupta D., et al., (2007). High Performance All Digital RF Receiver Tested at 7.5 GigaHertz, *Millitary Communication.*

[56] Gupta D., et al., (2007). Digital Channelizing Radio Frequency Receiver, *IEEE Trans. Appl. Supercond.* 17:430-437.

[57] Webber R., et al., (2007). Operation of superconducting digital receiver circuits on 2-Stage GM cryocooler, *Advances in Cryogenic Engineering*, 53:927-974.

[58] Kirichenko A., et al., (2005). Superconductor Digital Receiver Components, *IEEE Trans. Appl. Supercond.* 15:249-254.

[59] Mukhanov O., et al., (2002). Superconductor digital-RF tranceiver components, *presented at SDR 02 Technical Conference and Product Exposition, 2002, SDR Forum, and document is released as Report for Department of Energy, 2006, issued by HYPRES Inc, approved for public release, distribution unlimited.*

[60] Gupta D., et al., (2002). Benefits of superconductive Digital-RF transceiver technology to future wireless systems, *Proc. of SDR 02 Technical Conference and Product Exposition.*

[61] Fujimaki A., et al., (2001). Broad band software-defined radio receivers based on superconductive devices, *IEEE Trans. Appl. Supercond.* 11:318-321.

[62] Koshelets V., et al., (2009). Integrated superconducting receivers, *Supercond. Sci. Technol.* 13:R53-R9.

[63] Wikborg E., et al., (1999). RSFQ Front-end for a software radio receiver, *IEEE Trans. Appl. Supercond.* 9:3615-3618.

[64] Mukhanov O., (2011). *History of Superconductor Analog-to-Digital Converters*, in book H. Rogalla and P. Kes, Ed., Taylor & Francis, London, UK, pp. 440-458, ISBN: 978-1-4398-4946-0, e-book ISBN: 978-4398-5948-4.

[65] Brock D., (2001). Digital RF Architectures for JTRS-Compliant Base Stations, presentation at *MILCOM 2001, Session C19 (Joint Tactical Radio System Architecture)*, Paper #541, 30 Octorber 2001.

[66] Cook P., Superconductor Microelectronics and Security for Software Defined Radio, presentation.

[67] Mukhanov O., et al., (2014). Superconducting Quantum Arrays for Wideband Antennas and Low Noise Amplifiers, *presentation at International Microwave Symposium IEEE*, 1-6 June 2014, Tampa Bay, FL, MTT-S.

[68] Suzuki H., et al., (2011). Design and Demonstration of a 5-Bit Flash-Type SFQ A-D Converter Integrated With Error Correction and Interleaving Circuits, *IEEE Trans. Appl. Supercond.* 21:671-676.

[69] Inmadar A., et al., (2013). Flash ADC comparators and techniques for their evaluation, *IEEE Trans. Appl. Supercond.* 23:1400308.

[70] Inmadar A., et al., (2015). Design and evaluation of Flash ADC, *IEEE Trans. Appl. Supercond.* 25:1400205.

[71] Inamdar A., et al., (2007). Quarter-Rate Superconducting Modulater for Improved High Resolution Analog-to Digital Converter, *IEEE Trans. Appl. Supercond.* 17:446-450.

[72] Inmadar A., et al., (2009). Progress in Design of Improved High Dynamic Range Analog-to-Digital Converters, *IEEE Trans. Appl. Supercond.* 19:670-675.

[73] Sarwana S., et al., (2002). High-sensitivity high-resolution dual-function signal and time digitizer, *App. Phys. Lett.* 80:2023-2025.

[74] Kaplan S., et al., (2001). A prescaler circuit for a superconductive time-to-digital converter, *IEEE Trans. Appl. Supercond.* 11:513-517.

[75] Kirichenko A., et al., (2003). Multi-Channel time digitizing systems, *IEEE Trans. Appl. Supercond.* 13:454-458.

[76] Vernik I., (2008). Ultrasensitive wideband integrated spectrometer for chemical and biological Agent Detection, *Int. J. of High Speed Electronics*, 18:87-98.

[77] Mukhanov O., et al., (2008). Superconductor Digital-RF Receiver Systems, *IEICE Trans. Electron.* E91-C:306-315.

[78] Herr A., (2008). Superconducting Digital Signal Processor for Telecommunication, *IEEE/CSC & ESAS European superconductivity news forum*, 4.

[79] Buehler S., et al., (2016). Generation and distribution of Josephson junction clock, *IEEE Trans. Appl. Supercond.* 26:1300804.

[80] Kirichenko D., et al., (2005). High Quality on-chip long annular josephon junction clock source for digital superconducting electronics, *IEEE Trans. Appl. Supercond.* 15:296-299.

[81] Vernik I., et al., (2003). Two-Phase 50 GHz On-Chip Long Josephson Junction clock source, *IEEE Trans. Appl. Supercond.* 13:587-590.

[82] Tolpygo S., et al., (2007). 20 kA-cm^2 Process developments for superconducting integrated circuits with 80 GHz clock frequency, *IEEE Trans. Appl. Supercond.* 17:946-951.

[83] Brock D., et al., (2000). A 50 GHz monolithic RSFQ Digital Phase Locked Loop, *IEEE MTT-S International Microwave Symposium digest*, 1:353-356.

[84] Hopkins P. F., et al., (2019). RF waveform synthesizers with quantum-based voltage accuracy for communications metrology, *IEEE Trans. Appl. Supercond.* 29:1301105.

[85] Spooner A., et al., (1997). Superconducting Direct Digital Synthesizer, *IEEE Trans. Appl. Supercond.* 7:2270-2273.

[86] Donnelly Ch., et al., (2019). 1 GHz Waveform Synthesis with Josephson Junction Arrays, *IEEE Trans. Appl. Supercond.* 30:3.

[87] Karpov O. V., et al., (2002). A digital AC Waveform Synthesizer based on a Josephson Junction Chip with Binary Sections, *Instruments and Experimental Techniques*, 45:660-664.

[88] Donnely Ch., et al., (2018). Jitter Sensitivity Analysis of the superconducting Josephson arbitrary waveform synthesizer, *IEEE Trans. Microwave and Technique* 66:4898-4909.

[89] Benz S., et al., (2014). Pulse-bias electronics and techniques for a Josephson arbitrary waveform synthesizer, *IEEE Trans. Appl. Supercond.* 24:1400107.

[90] Inamdar A., et al., (2007). Digital Encoder for RF Transmit Waverform Synthesizer, *IEEE Trans. Appl. Supercond.* 17:556-550.

[91] Mukhanov O., et al., (2007). Superconductor Components Direct Synthesizer, *IEEE Trans. Appl. Supercond.* 17:416-421.

[92] Narahashi S., et al., (2006). Cryogenic reciver front-end with sharp skirt characteristics, *Supercond. Sci. Technol.* 19:S416-422.

[93] Kirichenko A., et al., (2007). RSFQ asynchronous switch matrix, *IEEE Trans. Appl. Supercond.* 17:505-507.

[94] Gupta D., et al., (2019). Digital Output Data Links From Superconductor Integrated Circuits, *IEEE Trans. Appl. Supercond.* 29:1303208.

[95] Lehmann A., (2014). Improved High-Speed Data Recorder for Superconducting Digital-RF Receivers, *IEEE Trans. Appl. Supercond.* 24:1300404.

[96] Ortlepp T., et al., (2009). Superconductor-to-Semiconductor Interface Circuit for High Data Rates, *IEEE Trans. Appl. Supercond.* 19:28-34.

[97] Weber R., et al., (2009). Ultra-low heat leak YBCO superconducting leads for cryoelectronic applications, *IEEE Trans. Appl. Supercond.* 19:999-1002.

[98] Webber R., et al., (2009). Evaluation of a 4K 4-Stage pulse tube cryocooler for superconducting electronics, *Cryocooler* 15:657-664.

[99] Inamdar A., et al., (2009). Superconducting switching amplifiers for high speed digital data links, *IEEE Trans. Appl. Supercond.* 19:1026-1033.

[100] Kadin A., et al., (2007). Current Leads and Optimized Thermal Packaging for superconducting systems on multistage cryocoolers, *IEEE Trans. Appl. Supercond.* 17:975-978.

[101] Gupta D., et al., (2013). Low-Power High-Speed Hybrid Temperature Heterogeneous Technology Digital Data Link, *IEEE Trans. Appl. Supercond.* 23:1701806.

[102] Engseth H., et al., (2007). Room temperature interface for RSFQ digital signal processor, *IEEE Trans. Appl. Supercond.* 17:979-982.

[103] Kataeva I., et al., (2007). Scalable Matrix multiplication with hybrid SMOS-RSFQ digital signal processor, *IEEE Trans. Appl. Supercond.* 17:486-489.

[104] Gupta D., et al., (2003). Integration of cryocooled superconducting analog-to-digital converters and SiGe output amplifiers, *IEEE Trans. Appl. Supercond.* 13:477-483.

[105] Gupta D., et al., (2001). High-speed interchip data transmission technology for superconducting multi-chip modules, *IEEE Trans. Appl. Supercond.* 11:731-734.

[106] Mukhanov O., et al., (1997). Josephson output interfaces for RSFQ circuits, *IEEE Trans. Appl. Supercond.* 7:2826-2831.

[107] Lehmann A., et al., (2013). Superconducting digital-RF receiver with high-speed data recording, *Superconductive Electronics Conference (ISEC), 2013 IEEE 14th International.*

[108] Lehmann A., et al., (2017). Embedded RSFQ Pseudorandom Binary Sequence Generator for Multichannel High-Speed Digital Data Link Testing and Synchronization, *IEEE Trans. Appl. Supercond.* 27:1301806.

[109] Ravindran P., et al., (2017). Energy Efficient Digital Data Link, *IEEE Trans. Appl. Supercond.* 27:1301105.

[110] Khamas, S. K., et al., (1988). High-T_c superconducting short dipole antenna, *Electronic Letters*, 24:460-461.

[111] Lancaster, M., et al., (1992). Supercooled and superconducting small-loop and dipole antennas, *IEE Proceedings-H*, 139:264-270.

[112] Wu Z., et al., (1989). High Tc superconducting small loop antenna, *Physica C*, 162-164:385-389.

[113] Lancaster, M., et al., (1992). Superconducting antennas, *Supercond. Sci. Technol.* 5:277-279.

[114] Hansen R. C., (2006). *Electrically small superdirective and superconducting antennas*, John Wiley & Sons, Inc., p168.

[115] Huang, Y., (1991). A high temperature superconductor superdirectional antenna array, *Physica C*, 180:267-271.

[116] Lancaster M., et al., (1993). YBCO thick film loop antenna and matching network, *IEEE Trans. Appl. Supercond.* 3:2903-2905.

[117] Ivrissimtzis L. P., et al., (1994). On the design and performance of electrically small printed thick film $YBa_2Cu_3O_7$ antennas, *IEEE Trans. Appl. Supercond.* 4:33-40.

[118] Walker G., et al., (1977). Superconducting superdirectional antenna array, *IEEE Trans. Antenna and Propagation*, AP-25 p885.

[119] Itoh K., et al., (1992). Two-element superconductive array antenna composed of high-Tc superconducting small helical antenna, *J. Superconductivity*, 5:485.

[120] Itoh K., et al., (1993). High-Tc superconducting small antennas, *IEEE Trans. Appl. Supercond.* 3:2836.

[121] Hansen R. C., (1990). Superconducting Antennas, *IEEE Trans. Aerospace and Electronic Letters*, 26:345.

[122] Hansen R. C., (1991). Antenna Application of Superconductors, *IEEE Trans. Microwave Theory and Technique*, 39:1508.

[123] Chaloupka H., et al., (1991). Miniaturized High-Temperature Superconductor Microstrip Patch Antenna, *IEEE Transactions on microwave theory and technique*, 39:1513-1521.

[124] Wang H., et al., (1999). Aperture-Coupled Thin Film Superconducting Meander Antennas, *IEEE Transactions on Antennas and Propagation*, 47:829-836.

[125] Floch J., et al., (1996). Experimental results on a 38GHz high temperature superconducting microstrip antenna, *Microwave and Optical Technology Letters*, 13:255.

[126] Chung Dong-Chul, (2001). Broadband HTS microstrip antennas for satellite communication, *IEEE Trans. Appl. Supercond.* 11:107-110.

[127] Ohshima S., (2000). High-temperature superconducting passive microwave devices, filters and antennas, *Supercond. Sci. Technol.* 13:103-108.

[128] Richard M., et al., (1993). Superconducting microstrip antennas: an experimental comparison of two feeding methods, *IEEE Transactions on Antennas and Propagation*, 41:967-974.

[129] Lancaster M., et al., (1998). Thin-Film HTS Planar Antennas, *IEEE Trans. Appl. Supercond.* 8:168-177.

[130] Ivrissimtzis L. P., et al., (1994). High-Gain Series Fed Printed Dipole Arrays made of high-Tc superconductors, *IEEE Transactions on Antennas and Propagation*, 42:1419.

[131] Ivrissimtzis L. P., et al., (1995). Supergain printed arrays of closely spaced dipoles made of thick film high-Tc

superconductors, *IEE Proc.-Microw. Antennas Propag.*, 142:26-34.

[132] Ivrissimtzis L. P., et al., (1995). A high gain YBCO antenna array with integrated feed and balun, *IEEE Trans. Appl. Supercond.* 5:3199-3202.

[133] Herd J. S., et al., (1993). Experimental results on a scanned beam microstrip antenna array with a proximity coupled YBCO feed network, *IEEE Trans. Appl. Supercond.* 3:2840-2843.

[134] Herd J. S., et al., (1996). Twenty-GHz broadband microstrip array with electromagnetically coupled high-Tc superconducting feed network, *IEEE Transactions on microwave theory and technique*, 44:1384-1389.

[135] Ali M., et al., (1999). Superconducting patch array antenna on both-side YBCO thin film for satellite communication, *IEEE Trans. Appl. Supercond.* 9:3077-3080.

[136] Morrow J., et al., (1999). Circularly polarized 20-GHz high-temperature superconducting microstrip antenna array, *IEEE Trans. Appl. Supercond.* 9:4725-4732.

[137] Zeng X., et al., (2017). A Circularly Polarized HTS Microstrip Antenna Array with Controllable Cryostat, *Hindawi International Journal of Antennas and Propagation.*

[138] Chung D., et al., (2003). Circularly Polarized HTS Microstrip Antenna Array, *IEEE Trans. Appl. Supercond.* 13:301-304.

[139] Chung D., et al., (2005). HTS microstrip antenna array for circular polarization with cryostat, *IEEE Trans. Appl. Supercond.* 15:1048-1051.

[140] Lewis L., et al., (1993). Performance of TlCaBaCuO 30 GHz 64 Element antenna array, *IEEE Trans. Appl. Supercond.* 3:2844-2847.

[141] Ohshima S., (2000). Fabrication and characterization of HTS antennas for satellite communication and security system, *Asia Pasific Microwave Conference*.

[142] Superconducting Devices, *broshure issed by CSIRO Inc*, Sydney, Australia.

[143] Du J., et al., (2008). Noise temperature and linewidth of high temperature-superconducting heterodyne gigahertz oscillators, *Appl. Phys. Lett.* 93:033507.

[144] Du J., et al., (2012). Self-pumped HTS Josephson heterodyne tunable mixer, *Supercond. Sci. Technol.* 25:025019.

[145] Du J., et al., (2009). Terahertz imaging at 77K, *Supercond. Sci. Technol.* 22:114001.

[146] Du J., et al., (2011). Terahertz and Millimetre Wave Imaging with a Broadband Josephson Detector Working above 77 K, *J. Infrared, Millimeter, and Terahertz Waves*, 32:681-90.

[147] Beggren S., et al., (2013). Development of 2D bi-SQUID arrays with high linearity, *IEEE Trans. Appl. Supercond.* 23:13255658.

[148] Keenan, Sh., et al., (2017). Large scale 2D SQIF arrays, *Presentation at EUCAS,* 18th September 2017.

[149] Kornev V., et al., (2010). Progress in high-linearity multi-element Josephson structures, *Physica C*, 470:886-889.

[150] Oppenlaender J., et al., (2000). Non-Φ0-periodic macroscopic quantum interference in 1D parallel Josephson junction arrays with unconventional grating structure, *Phys.Rev.B*, 63:024511.

[151] Kornev V. K. et al., (2016). A Guide to Active Antennas Based on Superconducting Quantum Arrays, *IEEE Transactions on applied superconductivity* 26:1400104.

[152] Kornev, et al., (2014). Superconducting Quantum Arrays for Broadband RF systems, *Journal of Physics: Conference Series* 507:042019.

[153] Kornev V., et al., (2014). Superconducting Quantum Arrays, *IEEE Trans. Appl. Supercond.* 24:1-6.

[154] Kornev V., et al., (2017). Broadband active electrically small superconducting antennas, *Supercond. Sci. Technol.* 30:0103001.

[155] Kornev V., et al., (2015). Output power and loading of superconducting quantum array, *IEEE Trans. Appl. Supercond.* 25:1602005.

[156] Kornev V., et al., (2008). How to build up the high linearity SQIF Structure, *Journal of Physics: Conference Series* 97:012010.

[157] Kornev V., et al., (2009). High linearity SQIF-like Josephson-Junction Structures, *IEEE Trans. Appl. Supercond.* 19:741-744.

[158] Kornev V., et al., (2009). Performance advantages and design issues of SQIF for microwave applications, *IEEE Trans. Appl. Supercond.* 19:916-919.

[159] Oppenlaender, et al., (2005). Superconducting quantum interference filters operated in commercial miniature cryocoolers, *IEEE Trans. Appl. Supercond.* 15:936-939.

[160] Caputo P., et al., (2006). Quadratic Mixing of Radio Frequency Signals using Superconducting Quantum Interference Filters, *App. Phys. Lett.* 89:062507.

[161] Caputo P., et al., (2007). Two tone response of radiofrequency signals using the voltage output of a Superconducting Quantum Interference Filter, *J. Superconductivity and Novel Magnetism*, 20:25-30.

[162] Krey S., et al., (1999). Highly sensitive magnetometers based on YBaCuO Josephson Junction arrays, *Appl. Phys. Lett.* 74:293.

[163] Oppenlaender J., et al., (2002). Highly sensitive magnetometers for absolute magnetic field measurements based on quantum interference filters, *Physica C*, 368:119-124.

[164] Schultze V., et al., (2003). High-Tc Superconducting Quantum Interference Filters for sensitive magnetometers, *IEEE Trans. Appl. Supercond.* 13:775-778.

[165] Schultze V., et al., (2003). Improved high-Tc superconducting quantum interference filters for sensitive magnetometry, *Supercond. Sci. Technol.* 16:1356-1360.

[166] Oppenlaender J., et al., (2003). Two dimensional superconducting quantum interference filters, *IEEE Trans. Appl. Supercond.* 13:771-774.

[167] Caputo P., et al., (2004). High-performance magnetic field sensor based on superconducting quantum interference filters, *App.Phys.Lett.* 85:1389-1391.

[168] Shadrin A., et al., (2008). Fraunhofer regime of operation for superconducting quantum interference filters, *App.Phys.Lett.* 93:262503.

[169] Kornev V., et al., (2011). Design and Experimental evaluation of SQIF arrays with linear voltage response, *IEEE Trans. Appl. Supercond.* 21:394-398.

[170] Kornev, V., et al., (2011). Linear Bi-SQUID Arrays for Electrically Small antennas, *IEEE Trans. Appl. Supercond.* 21:713-716.

[171] Kornev V., et al., (2012). Array designs for active electrically small superconductive antennas, *Physica C*, 479:119-122.

[172] Prokopenko G., et al., (2013). DC and RF measurements of serial Bi-SQUID arrays, *IEEE Trans. Appl. Supercond.* 23:1400607.

[173] Kornev V., et al., (2013), Active Electrically Small Antenna Based on Superconducting Quantum Array, *IEEE Trans. Appl. Supercond.* 23.

[174] Kornev, et al., (2014). Superconducting Quantum Arrays for Broadband RF systems, *Journal of Physics: Conference Series* 507:042019.

[175] Andrade M., et al., (2015). Detection of far field radio frequency signals by niobium superconducting quantum interference device array, *IEEE Trans. Appl. Supercond.* 25:1603005.

[176] Shesca B., et al., (2015). Flux-coherent series SQUID array magnetometers operating above 77 K with superior white flux noise than single-SQUID at 4.2K, *Appl. Phys. Lett.* 107:162602.

[177] Mitchel E., et al., (2016). 2D SQIF arrays using 20000 YBCO high Rn Josephson Junctions, *Supercond. Sci. Technol.* 20:S362-S366.

[178] P. Caputo, et al., (2007). Two tone response in superconducting quantum interference filters, *IEEE Trans. Appl. Supercond.* 17:722.

[179] Richard M., et al., (1992). Performance of a four-Element Ka-Band high temperature superconducting microstrip antenna, *IEEE Microwave and Guided Wave Letters*, 2:143.

[180] Luine, J., et al., (1999). Application of a dc SQUID array amplifier to an electrically small antenna, *IEEE Trans. Appl. Supercond.* 9:4141-4144.

[181] G. Prokopenko, et al., (2015). SQIF Antenna Measurement in Near Field, *IEEE Trans.* SQ-P24.

[182] P. Caputo, et al., (2005). Superconducting quantum interference filters as absolute magnetic field sensors, *IEEE Trans. Appl. Supercond.* 15:1044-1047.

[183] Prokopenko, G., et al., (2005). Study of a multi-channel RF amplifier based on DC SQUID for 3-5 GHz band, *IEEE Trans. Appl. Supercond.* 15:741-744.

[184] Prokopenko, G., et al., (2003). An HTS X-band dc SQUID based amplifier: Modeling and development concepts, *IEEE Trans. Appl. Supercond.* 13:1046-1049.

[185] Prokopenko, G., et al., (2003). Dynamic characteristics of S-band DC SQUID amplifier, *IEEE Trans. Appl. Supercond.* 13:1042-1045.

[186] Kalabukhov, A., et al., (2002). A high-Tc L-Band SQUID amplifier combined with superconductive thin-film filters, *Physica C*, 368:171-175.

[187] Prokopenko, G., et al., (2002). HTS dc SQUID based rf amplifier: development concept, *Physica C* 368:153-156.

[188] Prokopenko, G., et al., (1999). Two-stage S-band DC SQUID Amplifier, *IEEE Trans. Appl. Supercond.* 9:2902-2905.

[189] Shitov, S., at al., (1999). Superconducting Chip Receivers for Imaging Applications, *IEEE Trans. Appl. Supercond.* 9:3773-3776.

[190] Kosheletzs, V., et al., (2000). Integrated superconducting receivers, *Supercond. Sci. Technol.* 13:53-69.

[191] Reich, T., et al., (2006). *Development of a digital SQUID device for high sensitive measurement of widely varying magnetic fields, 51.* International Wissenschaftliches Kollaquium, Technische Universität Ilmenau, Septem. 11-15, 2006.

[192] Reich, T., et al., (2005). Digital SQUID sensor based on SFQ technique, *IEEE Trans. Appl. Supercond.* 15:304-307.

[193] Tsuga, Y., et al., (2013). Asynchronous Digital SQUID Magnetometer With on On-Chip Magnetic Feedback for Improvement of Magnetic Resolution, *IEEE Trans. Appl. Supercond.* 23.

[194] Myoren, H., et al., (2015). Design and Performance of Digital SQUID with Sub-Flux Quantum Resolution, *15th International Superconductive Electronics Conference (ISEC).*

[195] van Duuren, M. J., et al., (1997). The use of double relaxation_oscillation_SQUIDs as a sensor, *IEEE Trans. Appl. Supercond.* 7:3469-3472.

[196] Podt, M., et al., (2001). Design of a fast digital double relaxation oscillation SQUID, *IEEE Trans. Appl. Supercond.* 11:1235-1238.

[197] Myoren, H., et al., (2011). Design of Digital DROS with SFQ up-down counter for Wide Dynamic Operation Range, *IEEE Trans. Appl. Supercond.* 21:387-390.

[198] Myoren, H., et al., (2012). Design of Digital DROS for Wide Dynamic Operation Range with Sub Single Flux Quantum Resolution, *Physics Procedia* 36:121-126.

[199] Myoren, H., et al., (2017). Noise Properties of Digital SQUID Using Double Relaxation Oscillation SQUID Comparator With Relaxation Oscillation Resonant Circuit, *IEEE Trans. Appl. Supercond.* 27:1602205.

[200] Myoren, H., et al., (2019). Design of Digital Filter for Digital SQUID With Sub-Flux Quantum Feedback Resolution, *IEEE Trans. Appl. Supercond.* 29:1303004.

[201] Likharev, K. K., et al., (1991). RSFQ Logic -Memory Family - A new Josephson-Junction Technology for sub-terahertz-clock-frequency digital systems, *IEEE Trans. Appl. Supercond.* 1:3-28.

[202] Vernik I., et al., (2005). Integrated millimeter-submillimeter superconducting digital spectrometer, *IEEE Trans. Appl. Supercond.* 15:419.

[203] Vernik I., et al., (2008). Ultrasensitive wideband integrated spectrometer for chemical and biological agent detection, *International Journal of High Speed Electronics and Systems*, 18:87-89.

[204] Mei Bin, (1997). *Low-noise THz Niobium SIS mixers*, PhD. Thesis, California Institute of Technology.

[205] Belitsky V., et al., (1995). Theoretical and experimental studies of Nb-based tuning circuits for THz SIS mixers, Sixth International Symposium on Space Terahertz Technology, p87.

[206] Belitsky V., et al., (1995). Fourier Transform Spectrometer Studies (300-1000 GHz) of Nb-based Quasi-Optical SIS detectors, *IEEE Trans. Appl. Supercond.* 5:3445.

[207] Hong J. Sh., (2011). *Microstrip Filters for RF/Microwave Applications*, Second Ed., A John Wiley & Sons, Inc.

[208] Zhou L., et al., (2018). Design of Ultra-narrowband miniaturized high temperature superconducting bandpass filter, *Progress In Electrmagnetics Research Letters*, 76:105-111.

[209] Noel J., et al., (2003). A switched high-Tc superconductor microstrip resonator using a MEM switch, *Supercond. Sci. Technol.,* 16:1438-1441.

[210] Prophet E., et al., (2005). Highly-selective electronically-tunable cryogenic filters using monolitic discretely-switchable MEMS capacitor arrays, *IEEE Trans. Appl. Supercond.* 15:956-959.

[211] Tan C. Y., (2006). Planar tunable HTS microwave filter with patterned ferroelectric thin film, *Supercond. Sci. Technol.,* 19:212-216.

[212] Vargas J. M., et al., (2007). Design and Fabrication of Two SwitchSuperconducting Microstrip Hairpin Filters Using Series MEMS Switches, *IEEE Trans. Appl. Supercond.,* 17:898.

[213] Su H. T., et al., (2008). Novel tunable bandpass filter realized using barium-strontium-titanate thin films, *IEEE Trans. Microwave Theory & Technique*, 56:2468.

[214] Peng H., et al., (2009). Center frequency tuning for a narrowband superconducting filter in the time domain, *IEEE Trans. Appl. Supercond.,* 19:3683.

[215] Vendik I., et al., (2013). Tunable high-temperature superconductor filter using ferroelectric capacitors, *Proc. of 43 European Microwave Conference*, p672-675.

[216] Ying Zh., et al., (2013). Design and tuning of superconducting filter at VHF-Band with mechanically switchable interdigital capacitors, *IET Elecectronics Letters* 49:297-298.

[217] Wang J., et al., (2014). Varactor-tuned narrow-band high temperature superconducting notch filter, *IEEE Trans. Appl. Supercond.,* 24:01501104.

[218] Li Ch., et al., (2014). Tunable high-temperature superconductor bandpass filters realized using semiconductor varactors, *IEEE Trans. Appl. Supercond.,* 24:01501205.

[219] Suo G., et al., (2014). Low loss tunable superconducting dual-mode filter at L-Band using semiconductor varactors, *IEEE Trans. Microwave & Wireless Components Letters*, 24:170.

[220] Suo G., et al., (2014). Superconducting varactor tunable filter with constant bandwidth using coupling line, *IEEE Trans. Microwave & Wireless Components Letters*, 24:628.

[221] Jiang Y., et al., (2017). Superconducting tunable filter with constant bandwidth using coplanar waveguide resonators, *Chin. Phys. B*, 26:108501.

[222] Cui H., et al., (2019). Five order narrow band tunable superconducting filter, *In. J. of RF and Microwave Computer-Aided Engineering*, e.21820.

[223] Chaloupka H., et al., (1996). Superconducting planar disk resonators and filters with high power handling capability, *Electronic Letters* 32:1735-1737.

[224] Mansour R., et al., (1998). Design of high power superconductive output multiplexer, *IEEE Trans. MTT-S International Microwave Symposium Digest*, TH3A-3.

[225] Baumfalk A., et al., (1999). HTS power filters for output muliplexers in satellite communications, *IEEE Trans. Appl. Supercond.* 9:2857-2861.

[226] Schornstein S., et al., (1998). High-temperature superconductor-shielded high power dielectric dual-mode filter for applications in satellite communications, *IEEE Trans. MTT-S International Microwave Symposium Digest*, TH2A-4.

[227] Kolesov S., et al., (1997). Planar HTS structures for high power applications in communication systems, *J. of Superconductivity*, 10:179-187.

[228] Liu H., et al., (2019). *High-Temperature Superconducting Microwave Ceircuits and Applications*, Springer Singapore, ISBN: 978-981-13-6867-7.

[229] Martens J. S., et al., (1993). HTS-based switched filter banks and delay lines, *IEEE Trans. Appl. Supercond.* 3:2824.

[230] Pal S., (2006). Novel designs of high-temperature superconducting bandpass filters for future digital communication services, *J. Indian Inst. Sci.* 86:257-264.

[231] Shen Z., et al. (1997). Power Handling Capabiltiy Improvement of High-Temperatreu Superconducting Microwave Circuits, *IEEE Trans. Appl. Supercond.* 7:2446.

[232] Haiwen Liu, (2016). *Novel Compact High-Temperature Sueprconductor Multi-band Bandpass Filters*, Ph.D. work.

[233] Zheng, T., et al., (2004). Design of compact superconducting diplexer with spiral short-circuit stubs, *IEEE Trans. Appl. Supercond.* 24:1500405.

[234] Zhang Q., et al., (2010). A compact HTS duplexer for communication application, *IEEE Trans. Appl. Supercond.* 20:2-7.

[235] Feng C., at al., (2010). Design and performance of a compact superconducting duplexer for VHF-band, *Phys. C, Supercond.*, 470:1957-1961.

[236] Tao Y., et al., (2010). High isolation and compact diplexer using the hybrid resonators, *IEEE Microw. Wireless Compon. Lett,* 20:551-555.

[237] Heng Y., et al., (2013). Design and optimization of superconducting contiguous diplexer comprising doubly-terminated filters, *IEEE Trans. Appl. Supercond.* 23:1501706.

[238] Guan, X., et al., (2018). High isolation tubable diplexer based on mixed electromagnetic coupling, *Int. J. of RF and Microwave Computer-Aided Engineering*, 28:e21199.

[239] Guan X., et al., (2016). Miniaturized high-temperature superconducting diplexer usign common resonato and cross coupling structure, *IEEE Trans. Appl. Supercond.* 27:99.

[240] Guan X., et al., (2016). Compact, low insertion-loss and wide stopband HTS diplexer using novel coupling diagram and dissimilar spiral resonatros, *IEEE Tran. on Microwave Theory and Techniques*, 64:1-4.

[241] Kato T. et al., (2015). Power-handling Capability of superconducting filters using disk- and ring-type bulk resonators, *EEE Trans. Appl. Supercond.* 25:1501105.

[242] Zheng T., et al., (2015). Compact Superconducting Diplexer Design with conductor-backed coplanar waveguide structures, *IEEE Trans. Appl. Supercond.* 25.

[243] Wang X., et al., (2018). High isolation superconducting diplexer designed with double-side structure, *IEEE/MTT-S Int. Microwave Symposium.*

[244] Hong J.-Sh., et al., (2000). A high-temperature superconducting duplexer for cellular base-station applications, *IEEE Trans. on Microwave Theory and Technique*, 48:1336.

[245] Xu J., et al. (2013). A compact dueal-band bandpass superconducting filter using microstrip/CPW sprial resonators, *IEEE Microwave and Wireless Components Letters*.

[246] Huang F., et al., (2009). Dual-band superconducting sprial filters including narrow bandstop notches, IEEE *Trans. Microwave Theory and Techniques*, 57:1188-1195.

[247] Heng Y., et al., (2013). Dual-band superconducting bandpass filter with embedded resonator structure, *Electronics Letters*, 49:1096-1097.

[248] Hudrouss A., et al., (2008). A dual-band HTS filter using modified dual-spiral resonators, *IEEE Trans. Appl. Supercond*. 18.

[249] Ji L., et al., (2012). Design and performance of dula-nband high temperature superconducting filter, *Science in China Serid F, Information Sciences*, 55.

[250] Lu X., at al. (2018). High-order dual-band superconducting filter with independently controllable passbands and wide stopband, *Int. J. of RF and Microwave Computer-Aided Engineering*.

[251] Heng Y., et al., (2013). Dual-band superconducting bandpass filter using stub-loaded resonators with controllable coupling and feeding structures, *IEEE Microw Wirel. Compon. Lett*. 23:400-402.

[252] Lu X., et al., (2014). Design of a high-order dual-band superconducting filter with controllable frequencies and bandwidth, *IEEE Trans. Appl. Supercond*. 24:1-4.

[253] Song F. et al., (2015). Dual-band high-temperature superconducting bandpass filter using quint-mode stub-loaded resonators, *IEEE Trans. Appl. Supercond*. 25:1501410.

[254] Sekiya N., (2016). Design of high-order HTS dual-band bandpass filters with receiver subsystem fro future mobile communication systems, *Physica C*, 527:91-97.

[255] Sekiya N., et al., (2015). Design of miniaturized HTS dual-band bandpass filters using stub-loaded meander line resonators and their applications to tri-band bandpass filters, *IEEE Trans. Appl. Supercond*. 25:1500805.

[256] Liu H., et al., (2016). Compact dual-band HTS bandpass filter using spirally asymmetric stepped-impedance resonators, *IEEE Trans. Appl. Supercond*. 26:1501805.

[257] Liu H., et al., (2016). Dual-band high-temperature superconducting bandpass filter using dual-mode hairpin ring resonators, *IEEE Trans. Appl. Supercond.* 26:1501704.

[258] Liu H., et al., (2014). Dual-band superconducting bandpass filter using quadruple-mode resonators, *IEEE Trans. Appl. Supercond.* 24:130-133.

[259] Pond J. M., et al., (1991). Ultra-compact microwave filters using kinetic inductance microstrip, *IEEE Tran. Magnetics*, 27:2696.

[260] Zhang X., et al., (2012). Compact design of high-temperature superconducting dupleer and triplexer for satelllite communications, *Supercond. Sci. Technol.* 25.

[261] Wang J., et al., (2010). An L-band HTS duplexer with improved performance, *Chin. Sci. Bull.* 55:3527-3522.

[262] Heng Y., et al., (2014). A narrowband superconducting quadruplexer with high isolation, *IEEE Trans. Appl. Supercond.* 24:1500506.

[263] Song X, (2016). A superconducting manifold-coupled triplexer with an iterative design method at X-band, *Int. J. Microwave and Optical Technology Letters*.

[264] Liu H., et al., (2013). Triple-band high-temperature superconducting microstrip filter based on multimode split ring resonator, *App. Phys. Lett.* 103:142603-1-4.

[265] Liu H., et al., (2013). Dual-band superconducting bandpass filter using embedded split ring resonator, *IEEE Trans. Appl. Supercond.* 23:1300304.

[266] Liu H., et al., (2014). Compact high-temperature superconducting filter using multimode stub-loaded resonator, *IEEE Trans. Appl. Supercond.* 24:1500308.

[267] Song F., et al., (2016). A novel tri-band superconducting filter using embedded stub-loaded resonators, *IEEE Trans. Appl. Supercond.* 26:1502009.

[268] Liu H., et al., (2019). Novel Tri-band high-temperature superconducting bandpass filters using asymmetric shunted-line stepped-impedance resonator, *IEEE Acccess* 99.

[269] Liu H., et al., (2016). Tri-band high-temperature superconducting bandpass filters using multi-mode resonators, *IEEE Trans. Appl. Supercond.* 26.

[270] Zhou L., et al., (2018). Compact high temperature superconducting mulit-mode ultra-wideband filter, *Int. J. of Microwave and Optical Technology Letters*.

[271] Suo G., et al., (2014). Design of continously tunable superconducting filter with semiconductor varactors, *Int. J. of Microwave and Optical Technology Letters*, 56:775-777.

[272] Shojaei-Asanjan D., et al., (2019). A novel low-temperature superconductor power limiter, *IEEE Trans. Microw. Theory and Techniques*, 67:1005.

[273] Boot J., et al., (2004). Superconducting microwave power limiter for high-performance receiver protection, *IEEE MTT-S Int. Micrw. Symp. Dig.*, 1:139-142.

[274] Boot J., et al., (2003). A self-attenuating superconducting transmission line for use as a microwave power limiter, *IEEE Trans. Appl. Supercond.* 13:305-310.

[275] Rocas E., et al., (2011). Superconducting multiplexer filter bank for a frequency-selective power limiter, *IEEE Trans. Appl. Supercond.* 21.

[276] Rocas E., et al., (2011). Modelling of self-heating mechanism in the design of superconducting limiters, *IEEE Trans. Appl. Supercond.* 21.

[277] Belyaev B., et al., (2016). Reflective power limiter for X-band with HTSC switching element, *IEEE Trans. Appl. Supercond.* 26:1500506.

[278] Belyaev B., et al., (2012). Receiver protection device based on microstrip structure with high-temperature superconducting film, *Technical Physics Lett.*

[279] Govorun I., et al., (2018). Compact Microwave power limiter with HTSC element, Materials Science, *Journal of Siberian Federal University*.

[280] Govorun I., et al., (2017). HTS microwave power limiter based on microstrip quarter-wave resonators, *18th Int. Conf. of Young Specialists on Micro/Nanotechnologies and Electron Devices* (EDM).

[281] Sormeh Setoodeh, (2011). *Superconducting Microwave Filters*, Master Thesis, University of Waterloo, Canada.

[282] Prokopenko, G., et al., (2001). Low-noise S-band DC SQUID based Amplifier, *IEEE Trans. Appl. Supercond.* 11:1239.

[283] Prokopenko, G., et al., (1999). Two-stage S-band DC SQUID Amplifier, *IEEE Trans. Appl. Supercond.* 9:2902-2905.

[284] Prokopenko, G., et al., (1997). A dc SQUID based low-noise 4 GHz amplifier, *IEEE Trans. Appl. Supercond.* 7:3496.

[285] Mück M., et al., (2003). Superconducting quantum interference device amplifiers at gigahertz frequencies, *App. Phys. Lett*, 82:3266.

[286] DeFeo M. et al., (2012). Superconducting microstrip amplifer with sub-Kelvin noise temperature near 4 GHz, *App. Phys. Lett.* 101:052603.

[287] Zhang Q., et al., (2007). A HTS bandpass filter for a meteorogical radar system and its field tests, *IEEE Trans. Appl. Supercond.* 17:922.

[288] Keenan Sh., et al., (2013). High-Tc Superconducting Electronic Devices based on YBCO step-edge grain boundary junctions, *IEICE Trans. Electron.*, E96-C, 3: 298.

[289] Foley C., et al., (2020). A grain boundary Josephson Junction that supported may carrers and led to Applications with impact, *J. Superconductivity and Novel Magnetism.*

[290] Mitchell E., et al. (2019). Quantum interference effects in 1D parallel high-Tc SQUID arrays with finite inductance, *Supercond. Sci. Technol.* 32: 124002.

[291] Mueller, K. H., et al., (2021). Theoretical model for parallel SQUID arrays with fluxoid focusing, *Phys. Rev. B.* 103: 054509.

Chapter 3

OVERALL DESIGN OF L-BAND PHASED ARRAY ANTENNA SYSTEM BASED ON PARALLEL OPERATIONS OF SUPERCONDUCTOR AND SEMICONDUCTOR (CONVENTIONAL) ELECTRONICS

In this and the following chapters, an arbitrary phased array system, working in L-Band (1-2 GHz), is chosen just as an example in order to have some overview of possible applications of cryogenic receivers. The L-Band and possible further developments to S- and C-Bands are considered by the following reasons:

- L-, S-, C-Bands are widely used for RADAR systems,
- The same bands are also applied for the communication systems
- Some references of cryogenic receivers from the opening literature sources, see previous chapters and some references in this as well as in the following ones, also use L-Band.

The choice of the phased array is considered, while the single channel is a standard already well-known solution, and phased array systems have several additional advantages, like radar imaging, space resolution, etc.

So, an overall design and short summary of components, which are applied for the phased array system working with cold Low Noise Amplifiers (LNA) and cold super-/normal conductor components, are given. The cold antenna is considered as the first choice due to several reasons, i.e., easiness of connection between cold antenna and supporting structure, and ii) possibility to upgrade system (or to make further development) by applying an additional cold antenna, working on SQUID/SQIF principle.

Under the "conventional" system, we imply typical Cryogenic Receiver, i.e., one using superconducting filter, semi- or superconducting limiter, cold semiconductor LNA. Under the "superconducting" system, all other versions are implied, e.g.:

- Superconductor antenna, based on microstrip technology,
- Superconductor antenna, based on bi-SQUID/SQIF technology,
- Digital or analog SQUID/DROS antenna and corresponding electronics,
- RSFQ electronics.

It is foreseen that the "conventional" and "superconducting" systems are operated independently and have negligible influence on each other.

The goal of the present chapter is to give some overview on the cryogenic, vacuum and mechanical requirements and a realization of "supporting structure" (also named as "RF-Unit-Cell") for a single element of the phased array antenna. Such RF-Unit-Cells can be modified for holding different components, e.g., cold or warm antenna, simultaneous installation of different antenna types, installation of versatile electrical components, e.g., LNA, filter, RSFQ chips, etc. For that reason, only generic design is presented, and some more details on cryogenics will be also presented in the chapter 5.

3.1. RF Parameters: Specification

Main specification parameters could be summarized as follows (values are arbitrary chosen):

- Overall design concept is similar to one applied for the conventional cryogenic receivers. Main components are: cold or warm antenna, cold limiter (if necessary), superconducting filter, cryogenic LNA, and other warm components (other filters, ADC, etc.).
- Central frequency is 1.5 GHz (L-Band) and bandwidth 200 MHz, i.e., operation range 1.4-1.6 GHz.

The above-mentioned specification parameters are foreseen as for "conventional" as well as for "superconducting" systems. It is also possible that "superconducting" system will be later modified for operation at other frequency bands.

In the present chapter, the cryogenic receiver with warm antenna, superconducting filter and limiter as well as cold LNA is considered. Because antenna will be based on conventional design, i.e., without using of superconducting materials, it is not critical, whether it is cold or warm, because its efficiency is already very high. The situation is changed, if superconducting materials are used. The choice of superconducting antenna, which is based on conventional design, e.g., patch (i.e., without application of SQIF/SQUID elements), will be also shortly mentioned.

And finally yet importantly, it is worth particular to underline that for each component, the latest developments or further promising research activities will be considered. For that reason, only significant achievements over the last 15-20 years and promising developments have been taken into consideration.

3.2. RF Design: Advantages of "Conventional" Cryogenic Receiver

In this chapter, main advantages of "conventional" cryogenic receiver over completely "warm" receiver are considered and summarized. Under "conventional" cryogenic receiver, the receiver systems, which have been built in relatively large quantity and also mentioned in chapters 2.2 and 2.3, are implied. The cases of antenna with SQUID/SQIF elements and electronics working on Rapid Single Flux Quantum (RSFQ) logic will be considered in separate chapters. The case of "classical" antenna, e.g., patch, but made from superconducting materials, is also interesting and will be shortly mentioned. Such classical superconducting antennas were intensively investigated in 90's, but have been rarely applied for practical systems.

3.2.1. Superconducting Limiter

Application of sc limiter before LNA, allows reducing the losses from 3 dB to 0.5 dB or less (some advanced warm limiters have losses 0.5 dB or slightly below). It is also possible to avoid the limiter for the cryogenic receiver).

3.2.2. Superconducting Filter

Superconducting (sc) filter is applied in all cryogenic receivers. Enormous number of different sc filters has been developed over the last 30 years, and it is possible to find any design, which is suitable for each particular application, see chapters 2.2 and 2.8 for more discussion and practical examples. Extremely narrow bands, e.g., below 0.1%, and insertion losses of 0.5 dB or less, are "routine" experience. For non-tunable filters, any combination of low-, high-, band-pass (single or multiple), band-stop filters are possible.

For the present design, the filter will consist of two parallel branches and signal will be routed to any branch with active sc switch (an example of such switch could be found in Figure 2-17).

The first branch will be constructed from non-tunable sc filter with frequency range of 1.4-1.6 GHz. The losses are expected to be in the range of 0.1 dB. Just for the comparison, filters based on "conventional" technologies have following parameters:

- WBLB-T-BP-2500-200-5C (*A-Info* company) is cavity filter with insertion losses of 0.6 dB,
- WBLB-T-BP-2500-200-L (*A-Info* company) is Lumped Element filter with insertion losses of 3 dB,
- 885043 (*Qorvo* company) is BAW filter with insertion losses of 3.5 dB,
- ZFBP-2400+ (*Mini-Circuits* company) is microstrip filter with insertion losses of 3.5 dB.

Therefore, if filter before LNA is used, which is typically for cold but not the case for "warm" systems, an application of sc filter could theoretically give advantages of 0.5-2.9 dB, depending on each particular warm filter. The second branch will be constructed from tunable sc filter with tunability range of 1.4-1.6 GHz (8% relative) and as small as possible bandwidth, see Table 2-22 for listing of tunable filters. Tunable filters numbered as 5, 11-15 in Table 2-22, show that it is possible to make 8% tenability range and minimal fractional bandwidth could be as low as 1-2% (it could be even lower, see case "15" but insertion losses could be high). The tunable filters are under development, and further improvements, particular due to reduction of insertion losses and decreasing of relative bandwidth, will be possible.

Let us consider case "A," when after the LNA no reduction of bandwidth is foreseen, i.e., bandwidth of LNA is equal to the one of Analogue Digital Converter ("conventional" receiver). This case is more hypothetical because after LNA, bandwidth is reduced as much as needed. For "cryogenic" receiver, tunable filter is installed before LNA.

For the present conservative design for branch with sc tunable filter, it is assumed that the losses are around 0.7 dB, and relative bandwidth 0.5%. Due to reduction of the bandwidth, the noise level is decreased and system has improvement (sometime known as "dynamic gain") $10\cdot\log(BW_{wide}/BW_{narrow})$= 10□log(8%/0.5%)= 12 dB. In case if bandwidth of tunable sc filter is only 1%, the improvement is 9 dB. Therefore, for any case it is worth trying to reduce the bandwidth of the receiving system. Just for the comparison, non-tunable filter based on "conventional" technologies has the following parameters: ECA2492P24B90B (*ECHO Microwave* company), 2.480-2.50 GHz (0.8% relative) is cavity filter with insertion losses of 1 dB.

For the case "B" we consider that superconducting tunable filters are installed before LNA and after LNA, though after LNA it is not critical, whether the sc or warm filter is applied. So, in this case, the "warm" system with tunable filter after LNA, and "conventional" cryogenic receiver with the tunable sc filters before and after LNA are compared. Due to smaller bandwidth of sc filters, it is possible to reduce the bandwidth and effectively to suppress the noise. Assuming the bandwidth of warm tunable warm filter after warm LNA as 2% relative, the dynamic gain could be $10\log(BW_{warmReceiver}/BW_{coldReceiver})$= $10\log(2\%/0.5\%)$= 6 dB. So, in some cases, it is possible to find advantages of application of cryogenic receivers in comparison to "warm" ones. It is worth to note that this principle is very similar to the operation of Network Analyser, where inlet filter is used to reduce the bandwidth as much as possible.

3.2.3. LNA

Depending on each particular LNA design, see also chapter 3.4, performance improvement could be in the range of 0.5 till 0.8 dB. So, for LNA operating at 77K, the noise figure will be around 0.5-0.8 dB, while at room temperature – 1-1.6 dB. In order to be on the conservative side, it is assumed that cold LNA has 0.5 dB, and warm one – 0.8 dB.

3.2.4. *Matching* Network

Extensive measurements on the sc antenna, see chapter 2.6, showed that matching losses between antenna and transmission line play significant role, and in many cases they are the main contribution to the antenna losses. Application of sc materials allowed overall decrease of antenna losses including the matching one.

Due to the measurements on sc antenna properties, in many cases, it was quite difficult to separate the losses sources; so, to be on the conservative side, it is assumed that the improvement of system performance due to application of sc materials will be in the range 1-2 dB (or maybe up to 3 dB). It is worth to mention that these values are typically applied to the microstrip antennas, while for "volume" ones, the losses are small and typically dominated by losses at cable connections.

Recent trends in the development of communication systems showed that further improvements are possible and related to:

- Active impedance matching between antenna and other components
- active tuning of superconducting antenna, e.g., by aperture tuning with switch.

Therefore, it would be possible to further improve the receiver performance, though improvement "gain" will depend on each particular design modification.

3.2.5. Summary

For many cases, it is sufficient to make very rough comparison of possible advantages of cryogenic receivers over conventional ones. Table 3-1 shows possible advantages of receivers with cryogenic front end.

As one can note, even for the case if non tunable sc filter is used, the use superconducting components and cryogenic LNA could give some advantage.

And last but not the least, application of sc filters with very steep skirts and very high off-band rejection could bring another advantage of reducing of closed interference signals. This is actually the main reason, why sc receivers from *STI Inc* find their applications on market, see chapter 2.2.1 for more details.

Table 3-1. Extremely simplified comparison of cryogenic receivers with cold and warm ends

Parameter or component	Possible advantages	Comments
Superconducting limiter	0.5 till 3 dB	sc limiter has up to 0.5 dB losses. By using tunable filter with 1% bandwidth, it is possible to avoid the warm limiter.
LNA	0.3 till 0.8 dB	Warm LNA have 1-1.6 dB, while cold – typically 0.5-0.8 dB
Matching network	0.5 till 2 dB (maybe 3 dB or more)	Proper matching between antenna and transmitting line could be done with sc elements.
Superconducting Filter	-0.1 dB	For the estimations, it is assumed that sc filter is installed before LNA, while for warm system, the filter is absent
Total	1.2-5.7 dB	

3.3. Main Components and Estimation of Cold and Warm Receiver Noise Figures

Before one starts discussion of main components and their parameters, it is worth to compare the "cold" and "warm" receiver front ends with respect to their Noise Figure or effective temperature.

Table 3-2. Summary of components and their parameters used for estimations

Component	"Cold" front end	"Warm" front end	Comments
Matching network	0.0 dB (1.0)	0.0 dB (1.0)	
Filter	0.1 dB (1.0233)	0.0 dB (1.0)	Further gain due to reduction of bandwidth to 0.5% (or 1%) is separately considered, see pervious chapter.
Limiter	0.0 dB (1.0)	0.5 dB (1.1220)	No Limiter will be installed in the Cryogenic Receiver
LNA	0.5 dB (1.1220)	0.8 dB (1.2023)	

For the estimations, the efficiency losses in antenna and in matching network are neglected, and in "warm" receiver front end, the "virtual" band pass filter with zero losses is installed before LNA (however, in the most practical cases, it is installed after LNA). The physical temperature of cryogenic components is assumed to be around 60K and values are related to 1.4-2.1 GHz operation range. For the parameters of receiver with warm front end, the most optimistic values are chosen. Table 3-2 summarizes the parameters used for estimations.

The equivalent noise temperature of receiver with warm front end (without antenna physical temperature and with 100% efficiency) is:

$$T_{w,rec} = (L_{w,match} - 1) \cdot T_0 + L_{w,match} \cdot (L_{w,F} - 1) \cdot T_0 + L_{w,match} \cdot L_{w,F} \cdot (L_{w,Lim} - 1) \cdot T_0 + L_{w,match} \cdot L_{w,F} \cdot L_{w,Lim} \cdot (L_{w,LNA} - 1) \cdot T_0 = (1.1220 - 1) \cdot 290 + 1.11220 \cdot (1.2023 - 1) \cdot 290 = 35.4 + 65.8 = 101\,K \quad (1)$$

The equivalent noise temperature of receiver with cold front end (without antenna physical temperature and with 100% antenna efficiency) is:

$$T_{c,rec} = (L_{c,match} - 1) \cdot T_c + L_{c,match} \cdot (L_{c,F} - 1) \cdot T_c + L_{c,match} \cdot L_{c,F} \cdot (L_{c,Lim} - 1) \cdot T_c + L_{c,match} \cdot L_{c,F} \cdot L_{c,Lim} \cdot (L_{c,LNA} - 1) \cdot T_c = (1.0233 - 1) \cdot 60 + 1.0233 \cdot (1.1220 - 1) \cdot 60 = 1.40 + 7.49 = 9\,K \quad (2)$$

Assuming that antenna will be mostly directed towards the sky (vertical), the effective antenna temperature is conservatively assumed to be around 40 K (it could be lower to 4K, depending on properties of each supervising object), the effective noise temperature of warm and cold receivers are $T_{\text{warm,e}}$ = 101+40=141K and $T_{\text{cold,e}}$ = 9+40=49K. The corresponding Noise Figure, related to the 290K, are $NF_{\text{warm,e}}$ = 1+141/290=1.49 (1.72 dB) and $NF_{\text{cold,e}}$ = 1+49/290=1.17 (0.68 dB). So, sensitivity is increased by factor 1.27 (or 1.05 dB).

As one can note, for the system with best "warm" components, its Noise Figure could be only slightly behind than the one with cryogenic components. So, the main "gain" could be expected by the suppression of interferences or reduction of noise power due to application of tunable sc filter with small bandwidth.

One additional important conclusion it that application of “cold” (not superconducting) as well as superconducting antennas, which have very small losses, practically gives no further advantage, while antenna efficiencies are already close to maximum. In some case, due to reduction of losses, the antenna properties, like directivities, could be slightly varies, but such cases need detailed design consideration. So, it is important to consider also the effective temperature of the “source,” like “sky” or “fire,” because this temperature could dominate the receiver noise temperature.

3.4. Cold LNA

Figure 3-1 compares different MMIC (Monolithic Microwave Integrated Circuits) LNA.

It is possible to acquire the cryogenic LNAs from several suppliers, for example, *Low Noise Factory*, *Quantum Microwave*, *Callisto*, *MITEQ*, *Cosmic Microwave Technology*, *L-TEQ Microwave Technology*, etc. It would be possible to say that this technology reached a maturity, and further development is mainly related to the cost reduction, increasing of the LNA reliability after thermal cycling, minimizing of the thermal heat load, increasing of frequency range.

It is also worth to note that Fraunhofer IAF, Freiburg, is actively developing different room and cryogenic electronic components based on GaAs mHEMT (metamorphic High Electron Mobility Transistor) Technology.

Though all measurements and tests of LNA at cryogenic temperatures require wide experience on measurement technique and are very time consuming; technology maturity and large diversity of cryogenic LNA for L- and S-Bands provided from manufacturing firms or scientific institutions gives confidence on the successful application of cryogenic LNA and residual risks of the technology failure could be considered as minimal.

And the last but not the least; it is possible to gain “*in-house*” expertise by buying circuit boards and required components, i.e., resistors, capacitors, transistors, and to assembly the LNAs.

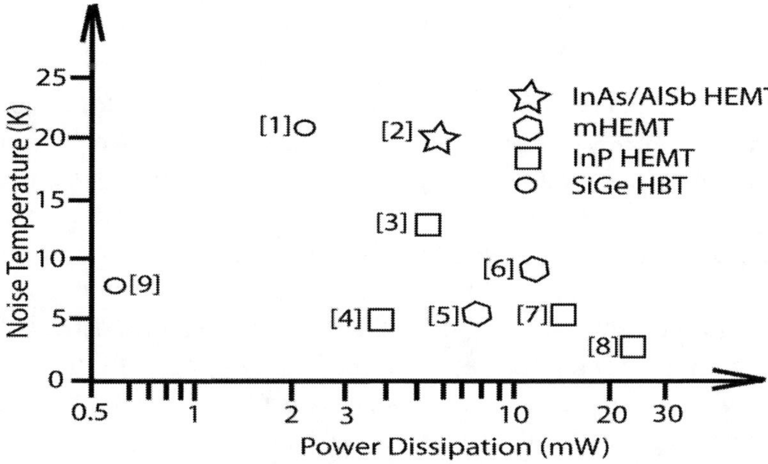

Figure 3-1. Comparison with *state-of-the-art* cryogenic MMIC LNAs [9]. References for this figure: [1] T. Tushar K., et al., *IEEE MWCL,* 18.7 (2008): 476-478. [2] G. M., et al. *IEEE MWCL* 22.3 (2012): 144-146. [3] R. Limacher *et al.*, *Indium Phosphide and Related Materials, 2004. 16th IPRM. 2004 International Conference on*. IEEE, 2004. [4] N. Wadefalk, et al. *IEEE MTT-S Int. Microw. Symp./Very Large Microw. Arrays Radio Astron. Space Commun. Workshop, Long Beach, CA*. 2005. [5] B. A. Abel´an *et al., IEEE TMTT* 60.12 (2012): 4080-4088. [6] B. Aja *et al.*, *IEEE MWCL* 21.11 (2011): 613-615. [7] J. Schleeh *et al.*, *IEEE TMTT* 61.2 (2013): 871-877. [8] N. Wadefalk, et al., *IEEE MTT-S Int. Microw. Symp./Very Large Microw. Arrays Radio Astron. Space Commun. Workshop, Long Beach, CA*. 2005, [9] Bardin J., (2017). Ultra-Low-Power Silicaon Germanium Cryogenic Low Noise Amplifiers, presentation given at *Cryogenic Engineering Conference*, 13 July 2017, Medison, USA, see also J. Bardin et al., (2017). Silicon Germanium Cryogenic Low Noise Amplifiers, *12th International Workshop on Low Temperature Electronics, IOP Conf. Series: Journal of Physics: Conf. Series* 834:012007.

3.5. Superconducting Filters

Depending on each particular design, it is possible to find or design suitable sc filter. In the time being, many sc filters have been developed by research institutions and, therefore, are customized for each particular application. However, several firms, like HYPRES Inc., STI Inc, Cryoelectra GmbH, Tianjin Hi-tech Developments Inc., have also some filters versions in their portfolio.

3.6. Cold Antenna

Though it is possible to realize cold or warm antenna, for the simplicity of design, it is preferable to use the cold one. This is related to the fact that LNA, which is operated at 5K temperature level, must be placed inside the "thermal radiation shield," which is kept at 50-80K temperature in order to reduce the thermal radiation heat load from the surrounding components. So, mechanical and thermal connection of antenna to the "shield" could be relative simple, while for the case of warm antenna, it has to be mechanically fixed on the thermal radiation shield but thermally has to be isolated. Moreover, application of the cold antenna gives a further advantage of reducing the signal losses in the cable between LNA and antenna, because in case of warm antenna, this cable must be long, e.g., 20 cm, in order to reduce a parasitic heat load over this cable to the cold LNA.

And the last but not the least; application of cold antenna, e.g., cavity for cold antenna, allows additional installation of other antennas, for example consisting of SQUID/SQIF blocks.

In this chapter, a short review of antennas, which are suitable for the effective cooling at the temperature range of 50-90K, as well as a design of the mechanical support structure for the antenna and LNA, are considered.

It is worth to mark that other versions of antenna supports are also possible, see the following chapters for more details.

3.6.1. Possible Choices of Cold Antenna

In order to effectively cool antenna and its single components, large contact surface areas between components inside antenna, as well as at connection place of antenna to the supporting structure at 50-80K temperature, must be foreseen. As a good example, the present GESTRA Cavity-Backed-Stacked-Patch-Antenna, could be considered [12]. Cavity has large contact surface at the bottom for good cooling, as well as large surface to the dielectric substrate. Due to large surface from substrate to the lower patch, the patch is also relatively good cooled. The upper patch, which is fixed on three long screws from isolating material, is relatively week cooled, but due to its lower mass, it is possible to keep it at temperature 20-30 K above the lower patch. Another good example could be micro-strip antenna manufactured on circuit boards. Due to large contact area between circuit board and supporting elements, as well as between micro-strips and dielectric material, it would be relative easy to keep it cold.

Table 3-3. Possible types of antennas applicable for cryogenic cooling

Type of antennas	Comments to cooling
Dipole, Monopole, Loops	Good, if placed horizontally to support, and bad if vertically
Microstrip	Good
Slot	Good, if manufactured on circuit board, which is horizontally mounted
Waveguide Slot	Good
Surface-wave	Good
Leaky-wave	Good
Helical	Bad
Frequency dependent (two- and four arms, log-periodic)	Good, if manufactured on circuit board, which is horizontally mounted
Horn	Bad
Reflector	Bad
Dielectric resonator	Good, if manufactured on circuit board, which is horizontally mounted
Lens	Bad
Conformal (spherical, 3D-dimensional)	Bad (good, only if surface is flat and mounted horizontally)
Vivaldy	Bad

In the Table 3-3 above, the antenna, which are good suitable for the cooling, are shortly summarized. Other antenna types are also shortly mentioned.

3.6.2. "RF-Unit-Cell" Design

Mechanical, cryogenic, high frequency (HF) and vacuum requirements on the "RF-Unit-Cell" (supporting structure for single element of the phased array antenna) could be summarized as follows:

Mechanical:

- Symmetric design in order to keep the antenna position in the middle.
- Compactness, for example, similar geometry to the GESTRA antenna.
- Small weight

Cryogenics:

- Operation temperature T<10K for LNA and temperature stability.
- Mechanical stability after thermal cycling between room and operational temperatures.
- Minimal shrinkage of components due to temperature decrease.
- Similar design in order to have possibility to scale-up to larger antenna array.
- “Modular” design in order to have possibility to scale-up to larger antenna array.
- If possible, design without fluids (gas or liquid phases), i.e., cooled with cryocoolers.

RF:

- Electrical connections must be done from the bottom (not from the side).
- Above the antenna, only materials with small RF-losses should be implemented.
- Above the antenna, only materials with measured dielectric constant and loss factor should be implemented.
- L- or S-Band frequency (this can be changed for other designs).

Vacuum:

- If possible, no materials with large surface areas, like foams, should be applied.

3.6.2.1. RF-Unit-Cell with cold antenna

Figure 3-2 shows the prototype of "RF-Unit-Cell" with warm Cavity Backed Stacked Patch Antenna. The design with cold antenna could be very similar.

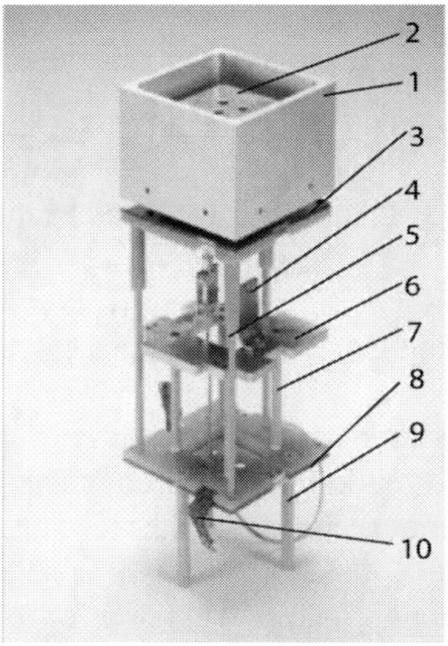

Main components:

- Cavity (marked as „1")
- Upper patch (marked as "2")
- Cu plate for the support of the antenna (marked as "3")
- Copper braid (marked as "10") for the connection to adjacent RF-Unit-Cells.
- Cu angle (marked as "4") used for support of LNA (LNA is not shown).
- G10 support rod (marked as "5") for plate with antenna
- 4K plate (marked as "6") from copper
- G10 foots for mechanic and thermal connection of 4K plate (marked as "7")
- Cu thermal radiation shield (marked as "8").
- G10 foots (marked as "9") for mounting of RF-Unit-Cell on vacuum vessel (vacuum vessel is not shown)

Figure 3-2. RF-Unit-Cell with warm Cavity Backed Stacked Patch Antenna [13, 14].

The present design of RF-Unit-Cell fulfills the requirements as follows:

- Mechanical stability is achieved by using a very robust and stable material G10 (other naming EP GC 201 or Hgw 2372).
- Components 9 and 7 are concentric to the center.
- Adjacent RF-Unit-Cells have similar geometry, so interconnection between them is possible.
- Temperature of LNA housing is below 10K

- G10 material is loaded with very low stresses.
- Thermal shrinkage of components is very small, and if necessary, it could be further reduced, e.g., by using Invar 36. No electrically conductive materials or ones with undefined dielectric constant or losses are placed above antenna.
- Electrical connectors are installed from bottom to top.
- No materials with large surface area are used.
- Cryogenic system uses no liquid or gaseous helium, i.e., it is cooled by cryocoolers.
- It is possible to install many RF-Unit-Cells and to increase the antenna diameter.

Due to sufficient space within RF-Unit-Cell it is also possible to install other chips, for example RSFQ one.

On the top of the cavity, it is possible to install the pick-up coils of flux transformer coils. Such coils could be made from superconducting materials and connected to other components, for example the digital receiver based on operation of RSFQ logic. Instead of pick-up coils, it is also possible to use other sc antennas, e.g., microstrip or bi-SQUID/SQIF ones.

3.7. Warm Antenna

The 3D model of RF-Unit-Cell with the warm antenna is shown in Figure 3-3 below. Main difference to the version with cold antenna is distance bolts (marked as "10"), which will be used to hold the reflector and antenna. As warm antenna type could be of any design, these distance bolts have arbitrary form and position.

The present designs of RF-Unit-Cells for warm or cold antenna are considered as "building blocks" for the construction of the large phased-array antennas.

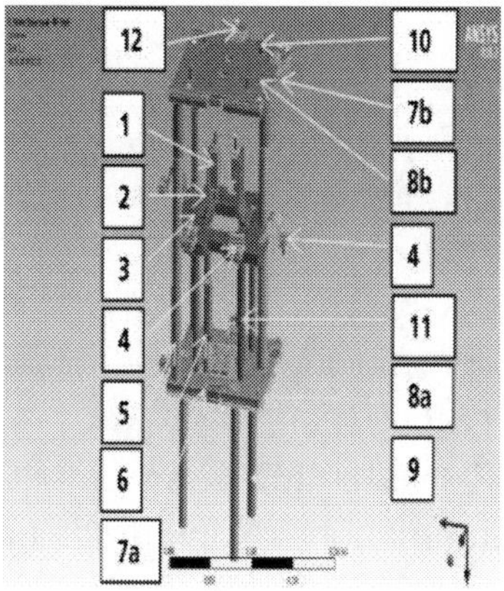

Main components:

- LNA (marked as „1“)
- Cu holder (“Angle”) for LNA (marked as “2”)
- 4K Cu Plate (marked as “3”)
- Cu braids for thermal connections between 4K plates (marked as “4”)
- G10 rod and distance holder (marked as “5”). Part 5 consists from two parts, which during the mounting are bolted together.
- Cu bar for thermal anchoring of RF-Cable (cable is not shown in figure) on 50-80K temperature (marked as “6”).
- Low (marked as “7a”) and upper (marked as “7b”) G10 plates
- Low (marked as “8a”) and upper (marked as “8b”) Cu plates, which are used as “thermal radiation shield” (temperature ca. 40-80K)
- G10 foots for mechanic and thermal connection of “RF-Unit-Cell” (marked as “9”) on vacuum vessel (vacuum vessel is not shown)
- G10 bolts (marked as “10”) for the holding of reflector and warm antenna (reflector and warm antenna are not shown).
- G10 foots (marked as “11”) for the holding of 4K Cu plate
- Cu braids (marked as “12”) for thermal connections between 80K Cu plates

Figure 3-3. Design of RF-Unit-Cell with warm antenna (warm antenna is not shown) [13, 14].

3.8. Comments to Chapter 3

In this chapter, the main specification parameters for “conventional” Cryogenic Receiver and “superconducting” one are mentioned. The application advantages of “conventional” receiver over “warm” one are considered.

The conceptual design of RF-Unit-Cell is also discussed. This design could be adapted for other variations of antenna (type, size, operational band) or cold electronic components.

For the similar design, the thermal and mechanical simulations were carried out [10, 11, 13]. Other research activities show that antenna array with 37 element RF-Unit-Cells and cold antenna is also feasible [11].

REFERENCES

[1] Thirivikraman T. K., et al., (2008). *SiGe HBT X-Band LNAs for Ultra-Low-Noise Cryogenic Receivers. IEEE Microwave and Wireless Components Letters,* 18(7):476-478.

[2] Moschetti G., et al. (2012). Cryogenic InAs/AlSb HEMT Wideband Low-Noise IF Amplifier for Ultro-Low-Power Applications, *IEEE Microwave and Wireless Components Letters*, 22(3):144-146.

[3] Limacher R., et al., (2004). 4-12 GHZ INP HEMT-based MMIC low-noise amplifier, 16th IPRM. 2004 International Conference on Indium Phosphide and Related Materials. p28-31; see also Robin F., et al. (2003). InP HEMT-based CPW amplifiers for 94-110 GHz, in: *International Conference on Indium Phosphide and Related Materials,* p110-13.

[4] Wadefalk N., et al., (2005). Very low noise amplifiers for very large arrays, *IEEE MTT-S Int. Microw. Symp./Very Large Microw. Arrays Radio Astron. Space Commun.* Workshop, Long Beach, CA.

[5] Abelan B. A., et al., (2012). 4-12 and 25-34 GHz cryogenic mHEMT MMIC low-noise amplifiers, *IEEE Trans. Microwave Theory and Technique,* 60(12):4080-4088.

[6] Aja B., et al., (2011). Cryogenic Low-Noise mHEMT-based MMIC Amplifiers for 4-12 GHz band, *IEEE Microwave and Wireless Components Letters,* 21(11):613-615.

[7] Schleeh J., et al., (2013). Cryogenic broadband ultra-low-noise MMIC LNAs for radio astronomy applications, *IEEE Trans. Microwave Theory and Technique,* 61(2):871-877.

[8] See reference [4] above.

[9] Bardin J., (2017). Ultra-Low-Power Silicaon Germanium Cryogenic Low Noise Amplifiers, presentation given at Cryogenic Engineering Conference, 13 July 2017, Medison, USA, see also J. Bardin et al., (2017). Silicon Germanium Cryogenic Low Noise Amplifiers, 12th International Workshop on Low Temperature Electronics, IOP Conf. Series: *Journal of Physics: Conf. Series* 834:012007.

[10] Ben-Bekhti N., et al., (2019). First studies towards a croy-cooled phased array radar system for space surveillance, *IOP*

Conference Series Material Science and Engineering, 502:012194.

[11] Barbri H., (2020). *Mechanische und thermische Simulation eines kryogenen 37-Element Phased-Arrays mit einer tiefsten Temperatur biz zu 4K zur Steigerung der Empfängersensitivität zur Detektion von Weltraumschrott*, Master Work, Hochschule für Technik und Wirschaft Berlin, 2020 in German [*Mechanical and thermal simulation of cryogenic 37-element phased array antenna with lowest temperature till 4K for increasing of receiver sensitivity for detection of space debries*.].

[12] Grenz O., (2017). *Verlustoptimierung einer breitbandigen, doppelt polarisierten CBSP-Antnne in L-Band mit Betrachtung der Kreuzpolarisationsentkopplung, Fertigungsaspekten und Phased-Array Tauglichkeit, Hochschule Bonn, Master Arbeit* [*Loss optimisation of wide-band double-polirized CBSP-antenna in L-Band with considereation of decoupling of mutual polarisation as well as production feasibility and phased-array application*].

[13] Froehlich A., et al., (2019). *Study towards a cryo-cooled Phased Array Radar System for Space Surveillance*, PAF Workshop, Bonn.

[14] Ben-Bekhti, N., (2018). *Jedes dB zählt – bessere Detektionsempfindlichkeit von Phased-Array-Radaren durch Kryotechnologie, in Fraunhofer FHR, Jahresberich*, in German, [*each dB is important – better detection sensitivity of phased-array-radars due to cryogenic technologies*].

Chapter 4

HF Design of Phased Array Antenna System Based on SC Electronics with Cold Antenna

In this chapter, some examples of application of the superconducting components for the design of receiver systems are given.

4.1. Case 1: Superconducting Filter and Cold LNA

In chapters 3.2 and 3.3, the main advantages of usage of superconducting components have been mentioned and could be summarized as follows: application of sc filter, limiter, matching network and semiconductor LNA could lead to sensitivity improvement of 1 dB. If narrow-band sc tunable filter is applied, the further improvement with 6 dB is possible.

It is worth also to mention that in the last 10-15 years, significant progress was done for the development of very compact wide range warm tunable antenna by using either i) active impedance matching between antenna and other components, as well as ii) active tuning of antenna, e.g., by aperture tuning with switch. Such developments could be also applied for the future radar systems.

4.2. Case 2: Superconducting Antenna, Filter and Cold LNA

This case is similar to the previous case, though an addition of antenna made from superconducting material give some advantages but also leads to the challenges on the impedance matching, see chapter 2.6 for more details. Advantage is an application of the superconducting materials for electrically small antenna, which have non-negligible electrical losses and impedance mismatch.

It is worth to underline that due to reduction of the electrical losses, sc antennas are narrowband and without active impedance tuning, it is quite difficult to use all advantages of the sc antennas. Some minor exception are radar or communication systems, which are design for very narrow frequency band and with very restricted overall dimensions, see for example chapter 2.3.5 for one such system.

For the above-mentioned reasons, the sc antenna could be applied, but only in the case if active impedance matching and active antenna tuning are realized.

In many cases, such antennas have small sizes, i.e., they belong to the classes of electrically small antenna, and constructed as planar microstrips. So, it is also possible to install them in parallel to other antennas, for example on cavities.

Based on available results in the opening literature sources, it could be possible that such antennas have gain improvement of 0.5-3 dB in comparison to the warm ones, see Chapter 2.6 for some references. For the more precise values, it is necessary to make thorough calculations for each particular sc antenna type.

4.3. Case 3: Electrically Small Active Antenna Based on bi-SQUID/SQIF Elements at 77K or 4K

In this chapter, a short rough design of superconducting quantum array antenna working at 4K and 40-80K temperature levels is presented, see also chapter 2.7 for more details on the design of bi-SQUID/SQIF antennas.

In order to limit discussion, it is assumed that present estimations will be done for signal frequency f_{signal}=2.5 GHz with bandwidth of *BW*=200 MHz. As our goal is to make rough estimations, only approximate calculations are presented. Detailed design could be done later for each particular system.

4.3.1. SQIF/bi-SQUID Antenna Working at 4K Temperature Level

Assumptions are: low-T_c superconducting junctions based on Nb technology with Josephson-junction normal resistance R_{N1}=1.2-1.6 Ohm (for calculations, the 1.4 Ohm is chosen) and supply current of I_c=0.125 mA per single Josephson Junction (JJ). The "unit," i.e., Differential Quantum Cell (DQS), will consist of two branches each with 10 parallel junctions, which are connected differentially over resistance of R_e. The Superconducting Quantum Array (SQA) will consist of two differentially connected arms, each consisting of K_p parallel blocks, which in turn further constructed of K_s serial differential quantum cells. Such design allows a variation of the dynamic range as well as facilitates the impedance matching.

- Differential Quantum Cell: Standard load impedance R_e=50 Ohm is chosen for easiness of impedance matching to the transfer line, e.g., coaxial or microstrip line. The parallel resistance of two arms each with 10 JJ is: R_N= R_{N1}/20=1.4/20=0.07 Ohm. Ratio R_e/R_n=50/0.07=714 and the characteristic voltage is V_c= R_{N1}□I_c =1.4Ohm□125μA=175 μV. Corresponding Josephson frequency f_j= V_c/Φ_0= 175□10^{-6}/2.07□10^{-15}=84.5 GHz.

  The dynamic range could be estimated as limit between V_c and root-mean-square voltage of the thermal noise.

$$V_T = \sqrt{2 \cdot 4k_BT\left(\frac{R_{N1}}{10}\right)BW} = \sqrt{2 \cdot 4 \cdot 1.38 \cdot 10^{-23} \cdot 5\left(\frac{1.4}{10}\right)0.2 \cdot 10^{9}} = 1.24 \cdot 10^{-7}V \quad , (3)$$

The dynamic range, which is also Spurious-Free Dynamic Range (SFDR) is:

$$DR_{singleDQC} = 20log\left(\frac{V_c}{V_T}\right) = 20log\left(\frac{175\ \mu V}{0.124\ \mu V}\right) = 63\ dB$$

Choice of parallel (K_p) blocks with serial (K_s) DQS.

The load impedance $R_{e,array}$ between two arms must be by factor 10 to 15 higher than total resistance of one arm of the superconducting quantum array. For the simplicity of impedance matching we also assume that $R_{e,array}$ is 50 Ohm, and the resistance of one arm should be: R_{arm}= $R_{e,array}$ /15=50/15= 3.33 Ohm.

The ratio serial to parallel block with DQS is: K_s/K_p= R_{arm}/ R_N=3.33 Ohm/0.07 Ohm=47.5≈48.

For the present technology development, around 600 DQS per single arm are manufactured. Taking this value as a guidance, the number of parallel blocks is estimated as K_p=3, and each block consists of K_s=200 DQS.

The dynamic range for array, which is also array SFDR is:

$$DR_{Array} = DR_{singleDQC}\sqrt{K_s \cdot K_p} = 20log\left(\sqrt{200 \cdot 3}\,\frac{175\ \mu V}{0.124\ \mu V}\right) = 90.8\ dB \quad (4)$$

Output characteristics: the output signal has a maximal differential voltage between two arms of array $\Delta V_{pp}=K_s \cdot V_c$=200$\cdot$175μV=35 mV. For the operation, it is assumed, that only α=80% of this voltage will be used, so the maximal voltage is V_{max}= $\alpha\Delta V_{pp}/2=0.8$$\cdot$0.035/2=14 mV. The maximal current amplitude is: I_{max}= $V_{max}/R_{e,array}$ =*14* mV /50 Ohm = 0.28 mA.

The maximal output power, P_{outMax}= $V_{max} \cdot I_{max}$ /2=1.96μW (-27 dBm).

The minimum voltage value, which is determined by thermal noise, is:

$$V_{\text{outMin}} = K_s \cdot \alpha \cdot \frac{V_c}{DR_{Array}} = 200 \cdot 0.8 \cdot \frac{0.000175}{90.8\ dB} = 0.81\ \mu V. \quad (5)$$

And minimal current, caused by thermal noise, is:

$$I_{\text{outMin}} = \frac{V_{\text{outMin}}}{R_{\text{e,array}}} = \frac{0.81 \cdot \mu V}{50\ Ohm} = 16.2\ nA. \tag{6}$$

Corresponding minimal power is:

$$P_{outMin} = V_{\text{outMin}} \cdot I_{\text{outMin}} = 0.81\ \mu V \cdot 16.2\ nA = 13.1 \cdot 10^{-15}\ W\ (-108\ dBm) \tag{7}$$

It is worth to note that calculated maximal and minimal values of the power, voltage and current are related to the measuring device. The voltage and current for one arm, which have to be supplied by a current source are: V_{source} = K_s□ V_c =200□175 μV=35 mV, I_{source} = K_p□*10*□I_c =3□10□125 μA=3.75 mA and corresponding averaged power P_{source} = V_{source} □ I_{source}=35mV□3.75 mA=131μW.

The field to voltage transfer function is taken as d*V*/d*T*=7000 mV/mT.

The maximal measured magnetic field:

$$B_{max} = \frac{V_{max}}{\frac{dV}{dT}} = \frac{14mV}{7000\frac{mV}{mT}} = 2\ \mu T \tag{8}$$

The minimal measured magnetic and electrical fields as well as magnetic field strength:

$$B_{min} = \frac{B_{max}}{DR_{Array}} = \frac{2\ \mu T}{90.8dB} = 58\ pT \tag{9}$$

$$H_{min} = \frac{B_{min}}{\mu_0} = \frac{58\ pT}{1.25 \cdot 10^{-6}\ H/m} = 46.4\ \mu A/m \tag{10}$$

$$E_{min} = H_{min} \cdot 377\ Ohm = 17.5 \cdot 10^{-3}\ V/m \tag{11}$$

Magnetic field noise:

$$\sqrt{S_{\rm B}} = B_{min}/\sqrt{BW} = 58 \cdot 10^{-12} T/\sqrt{0.2 \cdot 10^9} = 4\, fT/\sqrt{Hz} \qquad (12)$$

And corresponding voltage noise:

$$\sqrt{S_{\rm V}} = \sqrt{S_{\rm B}} \cdot {}^{dV}/_{dT} = 4 \cdot 10^{-15} T \cdot 7000 mV/mT = 28\, pV/\sqrt{Hz} \qquad (13)$$

It is worth to note that the minimal detectable magnetic field can be reduced by factor 10 till 100, if the superconducting washer (also known as “flux concentrator”) is used. For the read-out of signal ΔV_{pp} =35 mV it is possible to take a warm ADC with preamplifiers. In order to cover the large dynamic range of signals, it is possible to choose different warm pre-amplifiers with different minimal inlet signal sensitivities. Application of sc ADC for signal ranges -108/-27 dBm is also possible.

4.3.2. SQIF/bi-SQUID Antenna Working at 40-80K Temperature Level

Similar estimations are also done for antennas made from the high-T_c superconductors. Number of JJ and their parameters are taken from reference [1]. Only one arm is chosen, so no differential connection scheme is applied. The output resistance must be connected to the transfer line through impedance matching in order to reduce signal losses.

- Single Cell: The parallel resistance of single cell with 6 JJ is: $R_N = R_{N1}/6 = 1.9/6 = 0.32$ Ohm. Characteristic current strongly depends on the temperature, for example I_c=6, 13, 26 μA for T=43, 40, 35K, respectively; and for calculations is I_c= 10 μA for T=43K is chosen. Characteristic voltage is $V_c = R_{N1} \cdot I_c$ =1.9Ohm·10μA=19 μV. Corresponding Josephson frequency $f_j = V_c/\Phi_0 = 19 \cdot 10^{-6} / 2.07 \cdot 10^{-15} = 9.18$ GHz.
  The dynamic range could be estimated as limit between V_c and root-mean-square voltage of thermal noise.

$$V_T = \sqrt{2 \cdot 4 k_B T \left(\frac{R_N}{6}\right) BW} =$$
$$\sqrt{2 \cdot 4 \cdot 1.38 \cdot 10^{-23} \cdot 43 \left(\frac{1.9}{6}\right) 0.2 \cdot 10^9} = 5.48 \cdot 10^{-7} V \quad , \qquad (14)$$

The dynamic range, which is also Spurious-Free Dynamic Range (SFDR), is:

$$DR_{SingleCell} = 20 log\left(\frac{V_c}{V_T}\right) = 20 log\left(\frac{19\ \mu V}{0.548\ \mu V}\right) = 30.8\ dB \qquad (15)$$

Choice of serial (K_s) single cells.

In total K_s =5225 series cells are chosen. In original work [1], three different types with surface areas 16, 48, 80 μm^2 and corresponding inductances 11, 51 and 82 pH were used. The total resistance, R_{tot}= $K_s \cdot R_{N1}$=5225$\cdot$0.32 =1672 Ohm (measured value was ca. 1.4 kOhm).

The dynamic range for array, which is also array SFDR, is:

$$DR_{Array} = DR_{SingleCell}\sqrt{K_s} = 20 log\left(\sqrt{5225}\frac{19\ \mu V}{0.548\ \mu V}\right) = 68.0\ dB \qquad (16)$$

Output characteristics: the output signal has a maximal voltage ΔV_{pp}=$K_s \cdot V_c$=5225$\cdot$19 μV=99.3 mV. For the operation, it is assumed, that only α=80% of this voltage will be used, so the maximal voltage is V_{max}= α ΔV_{pp} */2=0.8*$\cdot$0.0993/2 =39.7 mV. The maximal current amplitude is: I_{max}= V_{max}/R_{tot} =*39.7* mV /1672 Ohm = 23.7 μA.

The maximal output power, P_{outMax}= $V_{max} \cdot I_{max}$ /2= 39.7$\cdot 10^{-3} \cdot$23.7$\cdot 10^{-6}$/2=0.47 μW or -33.2 dBm. This power must be supplied by the current source in order to keep SQIF structure at the voltage state.

The minimum voltage value, which is determined by thermal noise, is:

$$V_{\text{outMin}} = K_s \cdot \alpha \cdot \frac{V_c}{DR_{Array}} = 5225 \cdot 0.8 \cdot \frac{19 \cdot 10^{-6}}{DR_{Array}} = 31.7\ \mu V. \qquad (17)$$

And minimal current, caused by thermal noise, is:

$$I_{\text{outMin}} = \frac{V_{\text{outMin}}}{Rtot} = \frac{31.7 \cdot \mu V}{1672\ Ohm} = 19.2\ nA. \tag{18}$$

Corresponding minimal power is:

$$P_{outMin} = V_{\text{outMin}} \cdot \frac{I_{\text{outMin}}}{2} = 31.7\ \mu V \cdot 19.2 \frac{nA}{2} = 0.752\ \cdot 10^{-12}\ W = 0.304\ pW\ or - 95.2\ dBm \tag{19}$$

Connection to the further device: Depending on the type of device, which is connected to the output of this array, either V_{outMin} or I_{outmin} is relevant. If superconducting Analogue-to-Digital Converters (ADC) is connected to the outlet of array via transmission line with impedance *50* Ohm, the I_{outmin} is applicable in most of the cases. The ADC current sensitivity must be comparable with I_{outmin}.

Impedance matching: the impedance transformer for cable/microstrip lines, i.e., from 1.6 kOhm to 50 Ohm as well as from 50 Ohm to couple of ohms at ADC inlet must be designed.

The measured voltage at the linear region of SQIF is $V_{\text{max,Meas}}$= 11 mV, which is approximately factor 3 smaller than theoretical one V_{max}= 39.7 mV. Corresponding variation of magnetic field $B_{\text{max,Meas}}$= 12.5 μT. The field to voltage transfer function, d*T*/d*V*=12.5 μT/ 11 mV=1.14 μT/mV. The "voltage losses" due to real signal are: ΔV_{loss}=20□log($V_{\text{max,Meas}}$ /V_{max})= 20□log(11 /39.7)=11.1 dB.

The maximal measured magnetic field:

$$B_{max} = V_{max,Meas} \cdot \frac{dT}{dV} = 11mV \cdot 1.140 \cdot \frac{\mu T}{mV} = 12.5\ \mu T \tag{20}$$

The minimal measured magnetic field:

$$B_{min} = \frac{B_{max}}{DR_{Array} - \Delta\text{Vloss}} = \frac{12.5\ \mu T}{(68.0 - 11.1)dB} = 17.8\ nT \tag{21}$$

$$H_{min} = \frac{B_{min}}{\mu_0} = \frac{17.8\ nT}{1.25 \cdot 10^{-6}\ H/m} = 14.2\ mA/m \tag{22}$$

$$E_{min} = H_{min} \cdot 377\ Ohm = 5.4\ \ V/m \tag{23}$$

For the read-out of signal ΔV=99.3 mV it is possible to take the warm ADC with preamplifiers. In order to cover the large dynamic range of signals, it is possible to choose different warm pre-amplifiers with different minimal inlet signal sensitivities. Application of the sc ADC for signal ranges -98/-33 dBm is also possible.

The total area of superconducting SQIF loops is 0.11 mm^2 and no flux concentrator was applied. It is worth to mention that the minimal detectable signal could be lowered by factor 10 to 100, if the flux concentrator is applied (application of flux concentrator is typical). Concentrators have typical dimensions in the range of 2×2 till 10×10 mm^2.

Magnetic field noise:

$$\sqrt{S_{\mathrm{B}}} = B_{min}/\sqrt{BW} = 17.8 \cdot 10^{-9}T/\sqrt{0.2 \cdot 10^9} = 1.2pT/\sqrt{Hz} \tag{24}$$

And corresponding voltage noise:

$$\sqrt{S_{\mathrm{V}}} = \sqrt{S_{\mathrm{B}}} \cdot {}^{dV}\!/_{dT} = 1.2 \cdot 10^{-12}T \cdot 877\,V/T = 1.05\,nV/\sqrt{Hz} \tag{25}$$

In the following table, a summary of the most important parameters for the design of high-Tc SQIFs is presented (please note that many parameters were not published in the opening literature sources and the author had to estimate them).

One of the promising versions of high-Tc SQUID/SQIF antennas was realized by Ceraco Ceramic Coating GmbH and by groups at Loughborough and Nottingham Universities [2]. This design is also taken for the comparison. For low-Tc SQIF/bi-SQUID antennas, the most advanced versions were done by HYPRES Inc., though not many details are known on each design published in the opening literature sourced.

Table 4-1. Summary of the antenna parameters based on SQIF elements (only rough estimations)

Parameters	Present design	SCIRO (20,000 JJ)	SCIRO (100,200 JJ)	Ceraco Ceramic Coating GmbH
JJ resistance, R_n	1.9 Ω	5.34 Ω	5.34 Ω (assumed)	0.81 Ω
Critical current, I_c	10 μA at 43K	12.7 μA at 77K	30 μA at 77K	340 μA at 77K
$V_c=R_n \cdot I_c$	19 μV	67.8 μV	160 μV	275 μV
Number of parall. JJ	6	50	100	1 (only serial connection)
Resist. of parall. JJ	0.316 Ω	0.107 Ω	0.0534 Ω	0.81 Ω
Working Temp.	43K	77K	77K	77K
Bandwidth	0.2 GHz	0.2 GHz	0.2 GHz	0.2 GHz
RMS of thermal noise for parallel JJ.	0.548 μV	0.426 μV	0.301 μV	1.17 μV
SFDR for parallel JJ.	30.8 dB	44 dB	54.5 dB	47.4 dB
Number of serial blocks with parallel JJ	5225	400	1002	770
Resistance of array	1.65 kOhm	42.7 Ohm	53.5 Ohm	623 Ohm (measured 310 Ohm)
SFDR for array	64 dB	70.1 dB	84.5 dB	76.3 dB
Output ΔV(pp)	99.3 mV	27.1 mV	161 mV	10.1 mV (theoretical - 212 mV)
Max. output V(rms), for 80%	39.7 mV	10.9 mV	64.2 mV	4 mV
Max. output I(rms)	24 μA	254 μA	1.20 mA	6.5 μA (theoretical – 136 μA)
Maximal output power	-33.2 dBm	-28.6 dBm	-14.1 dBm	-48 dBm (theoretical – -22dBm)
Minimal output V(rms)	31.7 μV	6.82 μV	7.63 μV	26.1 μV
Minimal output I(rms)	19.2 nA	160 nA	143 nA	41.8 nA
Minimal output power	-95.2 dBm	-92.6 dBm	-92.6 dBm	-92.6dBm
dV/dT	877 V/T	1530 V/T	2480 V/T	(not available)
B(max)	12.5 μT	7.09 μT	2.58 μT	43 nT (from experimental values)
B(min)	17.8 nT	2.23 nT	0.154 nT	10 pT (from experimental values)
H(min)	14.2 mA/m	1.77 mA/m	0.122 mA/m	7.94 μA/m
E(min)	5.4 V/m	0.667 V/m	0.046 V/m	0.003 V/m

Parameters	**Present design**	**SCIRO (20,000 JJ)**	**SCIRO (100,200 JJ)**	**Ceraco Ceramic Coating GmbH**
Flux concentrator	No	No	No	Yes (integrated with each loop)
$S^{1/2}_B$	1.2 pT/Hz$^{1/2}$	0.158 pT/Hz$^{1/2}$	0.011 pT/Hz$^{1/2}$	0.467 fT/Hz$^{1/2}$
$S^{1/2}_V$	1.05 nV/Hz$^{1/2}$	0.24 nV/Hz$^{1/2}$	0.27 nV/Hz$^{1/2}$	
Surface area of single loop		280 μm^2	30 μm^2	1440 μm^2 (incl. flux concentrator)
Equiv. surface area of array		5.6×10^{-6} m^2	3.0×10^{-6} m^2	1.1×10^{-6} m^2
$S^{1/2}_{\Phi 0}$		426 μΦ_0/Hz$^{1/2}$	15.8 μΦ_0/Hz$^{1/2}$	0.25 μΦ_0/Hz$^{1/2}$ (experimental value) $S^{1/2}_V$ = 8 nV/ Hz$^{1/2}$.

4.4. Case 4: Digital SQUID and Digital DROS

For the present design estimations, the DROS parameters, which are presented in references [3-5], will be used.

Assumptions for the calculations: high-T_c superconductor for pick-up coil, low-T_c superconductor for digital SQUID/DROS and signal coils, neglecting inductances between pick-up and signal coils as well as parasitic capacitances. The pick-up coil has no correction due to the substrate material, has square form with d_p=113 mm (inner), and thickness of w_p=1mm.

The inductance and surface areas of the pick-up coil are $L_{pick\text{-}up} = 4.75\times10^{-7}$ H, and $A_{pick\text{-}up} = 1.3\times10^{-2}$ m^2.

The SQUID inductance is taken as L_{squid}=850 pH, number of turn is 50, the signal coil inductance $L_{signal}=50^2\times L_{squid} = 2.13\times10^{-6}$ H. The mutual inductance between SQUID and signal coils is $M_{signal}=0.8\times50\times L_{squid} = 3.4\times10^{-8}$ H. For the calculations, a strip inductance of the signal coil is neglected and 0.8 coefficient for the mutual inductance between SQUID and signal coils is chosen.

The magnetic flux must be "transferred" from the pick-up coil to the SQUID. The flux transfer factor describes which part of the signal is effectively coupled to the SQUID.

Transfer factor of the transformer coil (transformer coil is naming for the pick-up and signal coils) is:

$$K_{transfer\ factor} = M_{signal} \Big/ \left(L_{pick-up} + L_{signal}\right) = \frac{3.4 \cdot 10^{-8}}{4.75 \cdot 10^{-7} + 2.13 \cdot 10^{-6}} = 0.0131 \quad (26)$$

and effective area of the transformer coil is:

$$A_{pick-up,\ effective} = K_{transfer\ factor} \cdot A_{pick-up} = 0.0131 \cdot 1.29 \cdot 10^{-2} = 1.68 \cdot 10^{-4}\ m^2 \quad (27)$$

Clock frequency is taken as F_{clk}= $160 \cdot 10^9$ Hz. In the original paper [6], the 5 GHz clock frequency was used. For the present estimations, larger value is taken, which is related to the fact that due to significant research and development activities, significant work is done and higher values are surely expected, for example, HYPRES Inc has presently 100-1000 GHz clock in their portfolio.

Oversampling and sub-Φ_0 resolution ratios are taken as m=n=2^3=8. For the present signal with the center frequency of 2.5 GHz and BW=200 MHz, the Nyquist frequency $f_{Nyquist}$ = 5.5 GHz and 44 GHz sampling frequency will be sufficient for oversampling ratio 8.

Sub-Φ_0 resolutions, which is relevant for the feed-back loop and slew rate:

$$\Phi_{fb} = \Phi_0/2^n = 2.07 \cdot 10^{-15}/2^3 = 2.58 \cdot 10^{-16}\ V \cdot s\ (Wb) \quad (28)$$

Sub-Φ_0 resolutions, which is relevant for the magnetic flux resolution:

$$\Phi_{res} = \Phi_0/2^{n/2} = 2.07 \cdot 10^{-15}/2^{3/2} = 7.31 \cdot 10^{-16}\ Vs\ (Wb) \quad (29)$$

Flux noise or minimal detectable magnetic field flux signal.

$$\sqrt{S_\Phi} = \Phi_{res}/\sqrt{F_{clk}} = 7.31 \cdot 10^{-16}/\sqrt{160 \cdot 10^9} = 5.78 \cdot 10^{-21}\ Wb/\sqrt{Hz} \quad (30)$$

Magnetic field noise or minimal detectable magnetic field signal.

$$\sqrt{S_B} = \sqrt{S_\Phi}/A_{pick-up,\ effective} = 5.78 \cdot 10^{-21}/1.68 \cdot 10^{-4} = 3.43 \cdot 10^{-17}\ T/\sqrt{Hz} \qquad (31)$$

And minimal magnetic signal is:

$$B_{min} = \sqrt{S_B} \cdot BW = 3.43 \cdot 10^{-17}\sqrt{0.2 \cdot 10^9} = 4.85\ pT \qquad (32)$$

SQUID slew rate describes dynamical behavior of the measurement circuit.

$$\dot{\Phi}_{fb} = \Phi_{fb} \cdot F_{clk} = 2.58 \cdot 10^{-16} \cdot 160 \cdot 10^9 = 4.14 \cdot 10^{-6}\ V\ (Wb/s) \qquad (33)$$

And the slew rate related to the magnetic field:

$$\dot{B}_{fb} = \Phi_{fb}/A_{pick-up,\ effective} = 4.14 \cdot 10^{-6}/1.68 \cdot 10^{-4} = 2.46 \cdot 10^{-2}\ T/s \qquad (34)$$

For a sinusoidal magnetic field signal with the maximum amplitude B_0 and bandwidth *BW*, the slew rate must satisfy the following relation:

$$\dot{\Phi}_{fb} \geq 2\pi \cdot B_0 \cdot A_{pick-up,\ effective} \cdot BW \qquad (35)$$

The maximal amplitude of magnetic field is:

$$B_{0max} = \frac{\dot{\Phi}_{fb}}{2\pi \cdot A_{pick-up,\ effective} \cdot BW} = \frac{4.14 \cdot 10^{-6}}{2\pi \cdot 1.68 \cdot 10^{-4} \cdot 0.2 \cdot 10^9} = 19.5\ pT \qquad (36)$$

As one can note, the minimal and maximal values of magnetic field are nearly the same. In order to increase dynamic range, it is possible to reduce bandwidth, e.g., from 200 to 100 MHz, and in this case the field range is from 3.4 to 39 pT. So, it is possible to apply DROS with the loop antenna with a relative limited bandwidth and variation of absolute values of the detectable field could be performed by changing the geometry of the loop antenna and by decreasing of the bandwidth.

4.5. Case 5: Superconducting Filter and Antenna for High Power Operation

Due to experience on the development of sc filter for the high power applications, see chapter 2.8.5 for more details, it is possible to develop filter and antenna for the transmitter system with a high power operation level of around 100 W or more. Such transmitting system could be compact and applied as for a single as well as the phased array antennas.

4.6. Case 6: SQUID-Based Downconverter

Different mixers, which are based on operation of Josephson Junction or dc-SQUIDs, were developed for the frequency ranges up to 1 THz, see chapter 2.10.3 for some details. It could be possible that for some specialized cases, an application of the superconducting electronic elements will be justified and due to advances in SQUID manufacturing technologies, some more applications could be found in the future.

4.7. Case 7: Digital Receiver Based on HYPRES Inc. Development

Though quite significant progress and very impressive results in applications of the digital sc electronics have been achieved by HYPRES company, very limited information is available in the opening literature sources. For that reason, only data published in the scientific papers are evaluated. It is also worth to note that most of the data, which are published in the papers, are either preliminary or ones measured on the first prototype versions of devices, and, therefore, it will be naturally to expect that more advanced versions of devices (particularly ADC) are existing.

4.7.1. Digital Receiver with Warm Antenna

As it was mentioned in chapter 2.5, the digital receiver could be built for very different goals, e.g., after direct digitizing at 20-50 Gbps rate, signal can be either directly sent to warm electronics, or digitally filtered for required bandwidth, or auto correlated with the expected signal. Several different ADC with (or without) analogue filter before could be connected in parallel, and multiplexer could supply the digitized signals to the required rote.

Due to quite limited information available in the opening literature sources and challenges in direct requesting the HYPRES company for some basic information, only information presented in papers, presentation or brochures is taken for the following discussion.

In order to limit discussion, the simplified scheme is considered, i.e., sc analogue filter with center frequency 2.5 GHz and 200 MHz bandwidth, ADC (either flash or Δ-Σ types) and decimation filter with bandwidth of 200 MHz ($f_{Nyquist}$ =400 MHz).

Below some data, which will be used for estimations, are presented:

- Measured noise floor (power spectral density): -164 dBm/Hz, which corresponds to $S_{\Phi}^{1/2}$=2□10^{-21} T□m^2/Hz$^{1/2}$ [16, 19] and recalculated to 50 Ohm impedance $S_V^{1/2}$=2.2 nV/Hz$^{1/2}$.
- For the estimation of band-pass (ΣΔ) ADC, data presented in reference [17] will be taken. The signal frequency – f_{sig} =2.5 GHz, f_{clk} = 25.6 GHz, bandwidth – BW=200 MHz, SNR=43.6 dB, and SFDR=65 dB. These values correspond to the specification given at HYPRES web site, i.e., “up to 32 GHz Clock, 50-65 dB SFDR” and [24].
- Differential SQUID wheel comparator (DSW) showed 3.58 effective bits (22.2 dB) in binary code (4.58 in Gray) for 20 GHz input signal; see Figure 2-26 [18]. 24 transitions at 20 GHz showed digitization of a 480 GHz quantized current from quasi-

one junction quantizer. These values correspond to the specification given at HYPRES web site and [24], see also Table 2-6. For the further estimations, 5.0 effective bit (31.7 dB) at 2.5 GHz, see Figure 2-26, is taken.

- Another source of data is given in presentation [20]: i) 14-24 effective (true) bits, 100-160 dB Spur Free Dynamic Range, <100 kHz to > 2 GHz bandwidth, ii) ≈0K noise temperature, iii) Sensitivity -140 to -180 dBm, iv) Better than 10^{-15} bit error rate.

Below, the estimations of the flash and *Δ-Σ* ADC are presented.

For Flash ADC, the starting point is a clock frequency (which is actually the signal frequency, because some averaging is done before), 20 GHz, and 3.5 effective bits. Due to averaging in decimation filter, several additional bits will be synthetized in two parallel converters for *I* and *Q* components:

$$ENOB = log_2\left(\frac{m \cdot f_{clk}}{\pi \cdot f_{BandWidth}}\right) + 0.5 \cdot log_2\left(\frac{f_{clk}}{2 \cdot f_{BandWidth}}\right) = log_2\left(\frac{2 \cdot 20 \cdot 10^9}{\pi \cdot 200 \cdot 10^6}\right) + 0.5 \cdot log_2\left(\frac{20 \cdot 10^9}{2 \cdot 200 \cdot 10^6}\right) = 8.8\ bit \quad (37)$$

So, the total number of bits $N_{bitFlash}$= 3.5+8.8= 12.3 and SNR=76 dB.

For *Δ-Σ* ADC, due to a frequency down-conversion and additional averaging in decimation filter, additional bits will be also synthesized. For down-conversion, *I* and *Q* components will be used, each with defined decimation ratio.

$$ENOB = log_2\left(\frac{m \cdot f_{clk}}{\pi \cdot f_{BandWidth}}\right) + 0.5 \cdot log_2\left(\frac{f_{clk}}{2 \cdot f_{BandWidth}}\right) = log_2\left(\frac{2 \cdot 25.6 \cdot 10^9}{\pi \cdot 200 \cdot 10^6}\right) + 0.5 \cdot log_2\left(\frac{25.6 \cdot 10^9}{2 \cdot 200 \cdot 10^6}\right) = 9.3\ bit \quad (38)$$

So, the total number of bits $N_{\Sigma\Delta}$= (43.6-1.7)/6.02+9.3= 16.2 and SNR=43.6+57.7=101.3 dB. For further calculations, it is assumed that SFDR is also increased by 57.7 dB and will be $SFDR_{\Delta\Sigma}$ =65+57.7=122.7 dB.

It is also possible to apply some additional options, e.g.: i) multi-rate, ii) quarter-rate, iii) subranging. It is quite difficult to estimate, whether these options have been implemented or not, though most probably these options are already incorporated in new ADC versions.

So, application of sc ADC with the following digital filtering to the required bandwidth of ca. 200 MHz, the SNR and SFDR could be 101 and 122 dB, respectively. These values are better than ones of many semiconductor ADC.

Table 4-2. Main cryogenic components of digital RF-Receiver System (except Antenna)

Name	Parameters	Comments
8 ADC	Different bands types, e.g., band and low passes ones	Redundancy and spare ones are foreseen. Before each ADC, on-chip analog sc filter for required frequency and bandwidth will be installed
One Digital-RF Asynchronous Switch	8 inputs and 2 outputs	
Digital Channelizer	Down conversion, *I* and *Q* components, 8 parallel output channels for each component	Decimation ration of 2^8=256 is foreseen.
Decerializer	8 outputs after ADC	Rough data (without any operation on the digital data) after ADC will be transferred to the FPGA
Coaxial cables	16 for for digital channelizer; 16 (8 redundant) for Decerializer; 1 from room temperature clock, 1 from low temperature clock (after decimation), 1 for the signal testing.	UT-034 silver plated copper clad steel (also known as "silver-plated copper weld"). Estimated heat load at 4K temperature level is 0.4 mW. Expected data rate per single cable: 18-20 Gbps.
Chip dimensions	20*20 mm chip with 80 connections	Two chips will be applied, each for one polarization
Warm SiGe Amplifier	One per single coaxial cable	Optionally, differential amplification could be also applied, i.e., two coaxial cables per single "channel."

A possible configuration of the digital receiver is shortly discussed. It is worth to note that multi-functional, multi-channel systems have been developed by HYPRES Inc. for satellite communication and military applications, see Appendixes A and B.

In order to have a compact design of the cryogenics components, i.e., to reduce number of cables and thermal anchoring places for cables, relative low deserializing ratio, i.e., 8, is chosen (the same ratio is applied also for decimation filter).

For the flexibility of tests and operation, two outputs channels will be used, i.e., i) digital channelizer with ratio 8 for *I*and *Q* components, and, ii) decerializer with ratio 8. Number of the input channels for the Digital-RF Asynchronous Switch is chosen maximal in order to include 8 ADC. Table 4-2 summarizes chip parameter.

4.7.2. Digital Receiver with bi-SQUID/SQIF Antenna at 4K Temperature Level

HYPRES Inc. has developed electrically small active antenna based on the bi-SQUID block and operating at 4K temperature level, see chapter 2.7.2, Figure 2-41 and Table 2-16. In this case, the antenna was directly connected to the Advanced Digital Receiver (ADR), see chapter 2.5 for more details. The expected antenna sensitivity was in the range of $S_B^{1/2}$ = 50-200 fT/Hz$^{1/2}$ and $S_V^{1/2}$ = 15-60 μVm^{-1}/Hz$^{1/2}$, and for 200 MHz bandwidth, the magnetic and electric fields are B_{min} =0.7-2.8 nT, and E_{min} = 0.2-0.8 V/m. It is also worth to note that even smaller $S_B^{1/2}$ = 4.25 fT/Hz$^{1/2}$ were measured [21].

It is worth to underline that bi-SQUID/SQIF antenna has very low noise temperature around 1K at 1 GHz frequency, and a relative high amplification gain 20-40 dB [21-23]. For that reason, the total noise temperature of receiver could be practically neglected and only "sky" temperature (also called "antenna" one) will be measured.

HYPRES company has also developed the bi-SQUID/SQIF antenna working at 40-80K temperature level, though in this case, no ADR system was used. Unfortunately, nearly no information is available in the opening literature sources, and, therefore, it is difficult to make any estimations.

4.8. Case 8: Digital Transmitter (Direct Digital Synthesizer)

In the recent years, few firms and research institutions (mainly in USA) tried to develop several compact prototypes of digital transmitter [7-15].

It would be interesting to develop a digital transmitter module operating in L-Band (1-2 GHz).

The main parameters and components of the single-element transmitter module could be summarized as follows:

a) L-Band.
b) Sufficiently powerful Digital-to-Analog Converter (DAC) in order to supply signal to High Power Amplifier (HPA).
c) Application of digital predistortion scheme.

It is worth particular to stress, that though intensive developments in the field of the digital transmitters are done, no references for the functioning transmitters used for routine operation were found in the opening literature sources (only few references for small prototypes). It could be related to the fact that DAC operating in L-Band (unfortunately with relative low output power) was only recently developed by the research institutions in USA.

4.8.1. DAC

RSFQ dual binary-type digital-analog converter (DAC) is based on a combination of the binary digital-to-frequency converters (DFC) and binary summing SFQ arrays. Each SFQ pulse train from DFCs generates an output voltage across the corresponding serial stage of the SFQ summing array. The final sum of the induced voltages from all stages appears at the analog output.

It is worth to note that DAC operating in giga-Hertz region is still a challenge. Recent developments are based on application of Josephson Arbitrary Waveform Synthesizer (JAWS) done at NIST (National Institute of Standards and Technology, USA) and could be summarized as follows:

- Lumped element RF JAWS with 102 Josephson Junctions: 1 GHz sine wave synthesized over 10 MHz bandwidth. The power is -71 dBm, and SFDR is 66 dBc, 10 Hz resolution bandwidth.
- Diplexer RF JAWS circuit with 1440 Josephson Junctions: 1 GHz sine wave synthesized over 10 MHz bandwidth. The power is -49 dBm, and SFDR is 79 dBc (input digital stream was delivered at 4.8 GHz rate).
- SFQ JAWS with DC-to-SFQ converter: 4 GHz sine wave synthesized over 10 MHz bandwidth. The power is -74 dBm, and SFDR is 70 dBc.

4.8.2. Some Prototypes Built

One of the first superconducting Direct Digital Synthesizer (DDS) was designed by groups of TRW Space & Electronics Group and Sierra Monolithics in 1997. It included 12-stage Incremental Phase Accumulator, a 10-bit Sine-Function Generator (SineROM) and 8-bit Digital-to-Analog Converter. The demonstrated parameters could be summarized as follows: instantaneous bandwidth – 500 MHz, frequency resolution 500 kHz, spurious content – -30 dB, clock rate 2 GHz, operation frequency (goal) – 1 GHz, and output power from DAC – -60 dBm.

HYPRES Inc has also demonstrated operation of single components [9].

Unfortunately, no further references have been found in the opening literature sources.

4.9. Case 9: SFQ-Based Computer

Since 1980s (mainly in Japan and USA), activities towards developments of superconducting computers based on operation of the RSFQ logic elements are undertaken.

Two main goals are:

- To develop alternative to the present SMOS technology,
- To develop Quantum Computer.

Due to limited information presented in the opening literature sources, only activities in USA under IARPA program "Cryogenic Computing Complexity" were notices. The main goal is a development of the 64-bit computing machine operating at 10 GHz clock frequency with a throughput of 10^{13} bit-op/s. Similar programs are also on-going in China and Japan, though little information is available on the status and goals of projects.

References

[1] Cybart S., et al., (2014). Large voltage modulation in magnetic field sensors from two-dimensional arrays of YBaCuO nano Josepshon junctions, *Appl. Phys. Lett.* 104:062601.

[2] Shesca B., et al., (2015). Flux-coherent series SQUID array magnetometers operating above 77 K with superior white flux noise than single-SQUID at 4.2K, *Appl. Phys. Lett.* 107:162602.

[3] Podt, M., et al., (2001). Design of a fast digital double relaxation oscillation SQUID, *IEEE Trans. Appl. Supercond.* 11:1235-1238.

[4] Myoren, H., et al., (2011). Design of Digital DROS with SFQ up-down counter for Wide Dynamic Operation Range, *IEEE Trans. Appl. Supercond.* 21:387-390.

[5] Myoren, H., et al., (2012). Design of Digital DROS for Wide Dynamic Operation Range with Sub Single Flux Quantum Resolution, *Physics Procedia* 36:121-126.

[6] Myoren, H., et al., (2019). Design of Digital Filter for Digital SQUID With Sub-Flux Quantum Feedback Resolution, *IEEE Trans. Appl. Supercond.* 29:1303004.

[7] Mukhanov O., et al., (2002). Superconductor Digital-RF Transceiver Components, *presented at SDR 02 Technical Conference and Product Exposition, SDR Forum,* and document is released as *Report for Department of Energy, 2006, issued by HYPRES Inc, approved for public release, distribution unlimited.*

[8] Gupta, D., et al., (2002). Benbefits of superconductor Digital-RF Tranceiver Technology to Future Wireless Systems, *Proceedings of SDR 02 Technical Conference and Product Exposition*, SDR Forum.

[9] Mukhanov O., et al., (2007). Superconductor Components Direct Synthesizer, *IEEE Trans. Appl. Supercond.* 17:416-421.

[10] Hopkins P. F., et al., (2019). RF waveform synthesizers with quantum-based voltage accuracy for communications metrology, *IEEE Trans. Appl. Supercond.* 29:1301105.

[11] Spooner A., et al., (1997). Superconducting Direct Digital Synthesizer, *IEEE Trans. Appl. Supercond.* 7:2270-2273.

[12] Donnelly Ch., et al., (2019). 1 GHz Waveform Synthesis with Josephson Junction Arrays, *IEEE Trans. Appl. Supercond.* 27.

[13] Karpov O. V., et al., (2002). A digital AC Waveform Synthesizer based on a Josephson Junction Chip with Binary Sections, *Instruments and Experimental Techniques*, 45:660-664.

[14] Donnely Ch., et al., (2018). Jitter Sensitivity Analysis of the superconducting Josephson arbitrary waveform synthesizer, *IEEE Trans. Microwave and Technique* 66:4898-4909.

[15] Benz S., et al., (2014). Pulse-bias electronics and techniques for a Josephson arbitrary waveform synthesizer, *IEEE Trans. Appl. Supercond.* 24:1400107.

[16] Gupta D., et al., (2014). Multi input Synchronous Analog to Digital Converter, *IEEE/CSC Superconductivity News Forum, 2014, IEEE Trans. Appl. Supercond.* 24:1400107.

[17] Gupta D., et al., (2011). Superconductor Analog-to-Digital Converters and Their Applications, *IEEE MTT-S International Microwave Symposium Digest*, 12240705.

[18] Inmadar A., et al., (2015). Design and evaluation of Flash ADC, *IEEE Trans. Appl. Supercond.* 25:1400205.

[19] Radparvar M., (2015). *Digital SQUID Magnetometers for read-out of detectors and magnetic particles*, August 7, 2015, presentation.

[20] Cook P., *Superconductor Microelectronics and Security for Software Defined Radio*, presentation.

[21] Berggren S., (2013). Development of 2D Bi-SQUID Arrays with high linearity, *IEEE Trans. Appl. Supercond.* 23.

[22] Mukhanov O., et al., (2014). Superconducting Quantum Arrays for Wideband Antennas and Low Noise Amplifiers, *presentation at International Microwave Symposium IEEE*, 1-6 June 2014, Tampa Bay, FL, MTT-S.

[23] Prokopenko G., et al., (2013). DC and RF measurements of serial Bi-SQUID arrays, *IEEE Trans. Appl. Supercond.* 23:1400607.

[24] HYPRES Brochure (2015) *World-Class Foundry Services*.

Chapter 5

MECHANICAL, CRYOGENIC AND VACUUM DESIGNS OF RECEIVER

In this chapter, a short design summary of the whole cryogenic systems for the phased-array antennas with different number of RF-Unit-Cells is given.

5.1. REQUIREMENTS

Two main requirements on the cryogenic system are [1, 2]:

- To keep bottom of the vacuum vessel free from any mechanical components in order to have space for the installation of electronic components.
- To have reasonable number of cryocoolers for large arrays, e.g., system with >100 RF-Unit-Cells.

5.2. MODULAR CRYOSTATS WITH SEPARATE VACUUM VESSEL AND CRYOCOOLER

There are two possibilities to design the cryostat for the antenna: i) one single vacuum vessel, which includes a large number of the single antennas, and ii) small vacuum vessel, which has only limited number of antennas, e.g., 16, see Figure 5-2 and Figure 5-3. With the second method, it is possible to extend such structure to larger system, see Figure 5-4. For the RF-Unit-Cell, which is presented in the previous chapters, the first method could be easily applied, while in the following chapters– the second method is considered.

5.2.1. Modular Design of Cold Antenna and Electronics at 40-80K Temperature Level

The design with cold antenna and cold electronics operated at 40-80K temperature level could be also advantageous for two reasons, i.e., i) cold LNA, or ii) cold electronic components, e.g., ADC, filter, and electrically small broadband antennas based on SQUID/SQIF and manufactured from high temperature (high-Tc) superconductors.

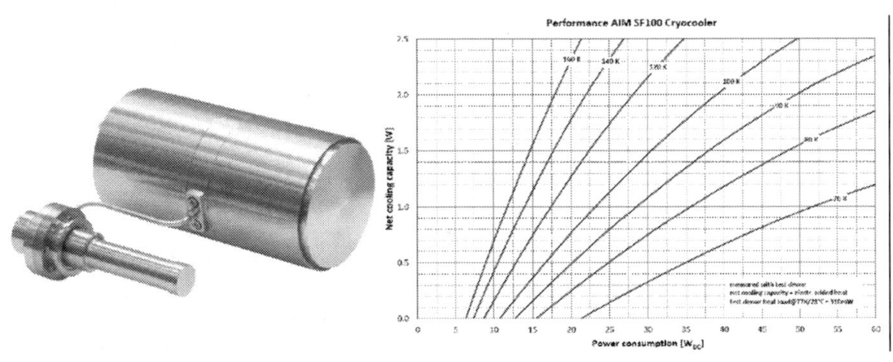

Figure 5-1. Bild and Load Map of SF100 Flecture Bearing Stirling cryocooler.

The main cryogenic component is small one-stage cryocooler, which are typically used for cooling of small objects, see Figure 5-1 for working map of SF100 Flexture Bearing Stirling cryocooler with cooling power of 2.0W @ 80K and electrical power consumption of 60W. Such cryocooler has MTTF >25,000 h, which allows maintenance free operation for 3 years and could be supplied with an integrated electronics (other option – external electronic unit for cryocooler control). The diameter and length of SF100A compressor are 60 and

120 mm, respectively, while the diameter and length of cold finger are 14.1 and 57.73 mm.

The modular design of the phased array system, which is based on the separate vacuum and cryogenic systems, i.e., one cryocooler and one vacuum vessel, is presented in Figure 5-2 below. After assembling and testing, it is possible to insert such system into the large supporting frame, see Figure 5-4.

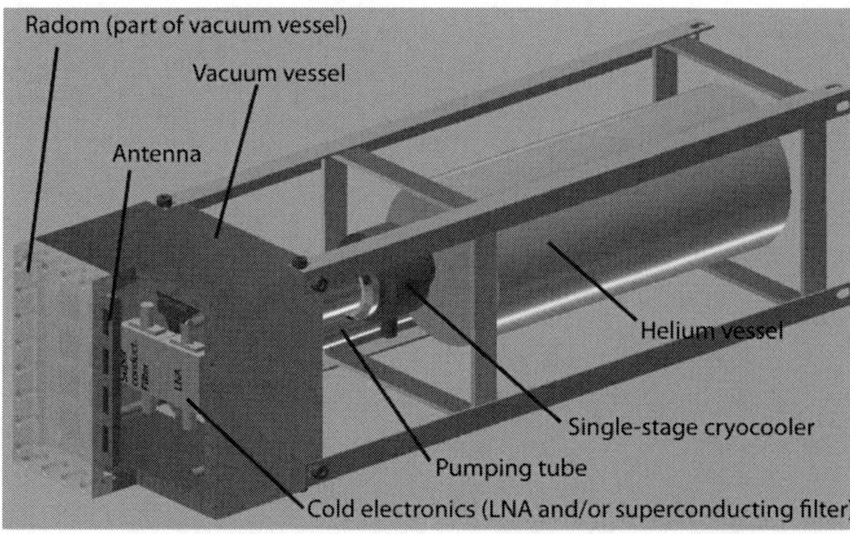

Figure 5-2. Modular design of RF-Unit-Cell with separate cooling unit and vacuum vessel. Operational temperature of LNA, filter and antenna are 40-80K.

The loop or patch antennas are particularly suitable for the applications, because they have large surface areas, which facilitates cooling, as well as the impedance matching for microstip transmission lines is relative easy. In this design, only a single stage small cryocooler with the operational temperature of 40-80K is used. Cold electronics includes sc filter, LNA, limiter.

It is worth to note, that it is possible to apply two "parallel" phased arrays, one is based on design with superconducting high-Tc material (most probably YBaCuO crystal), and another one - on copper (due to cryocooling, the losses in Cu will be reduced, which could also improve performance of the system).

The RF-Unit-Cell has dimensions of 84×84×300 mm, and if it is necessary, it is possible to increase the height of the vacuum vessel in order to install more cold electronic components.

The pumping tube is used for the initial evacuation of the vacuum vessel, and in the normal operation mode will not be used (it will be closed with small valve, which is located behind the helium vessel, see Figure 5-2). Such modular design has been already realized, see chapter 2.3.4 for more details.

5.2.2. Modular Design of Cold Antenna and Electronics at 4 K Temperature Level

The modular design of RF-Unit-Cell with antenna operating at 4K temperature level is shown in Figure 5-3. The 1.5W at 4.2K Sumitomo cryocooler RDK415D is foreseen for cooling. The module dimensions are 400×220×590 mm and mainly limited by sizes of the cryocooler.

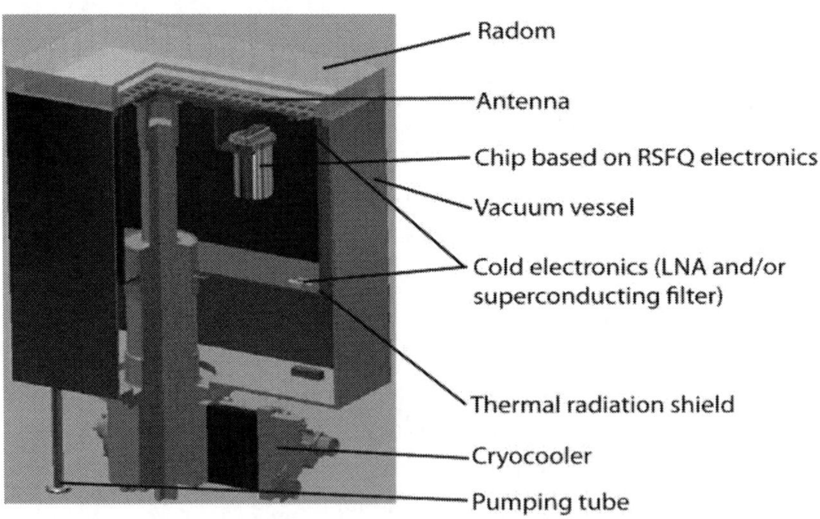

Figure 5-3. Modular design of RF-Unit-Cell with 4.5K temperature of cold electronics and antenna.

This module will be inserted into the supporting structure, see Figure 5-4 below. It will be possible to further extend the phased array antenna by an installation of additional supporting frames with modules.

Figure 5-4. Supporting frame for a modular design of the phased array system. Each module has separate cryocooler and insulation vacuum vessel and can be operated either at 40-80K temperature level by using single-stage cryocooler, or at 4K and 40K levels by using double-stage one.

5.3. Comments and Notes to Chapter 5

Though cryogenics is still considered as a major handicap for many applications (not only radars and communication systems), many diverse solutions have been developed practically for near all engineering cases.

Depending on each particular case, the cryogenic system will have some specific design requirements, which are surely realizable, though sometimes the realization will be time, manpower and costs intensive. In many cases, the work to be done is underestimated.

For the future radars and communication systems, it is possible to say, that if components, which require low temperatures, are needed, the cryogenic system is also realizable.

References

[1] Ben-Bekhti N., et al., (2019). First studies towards a cryo-cooled phased array radar system for space surveillance, *IOP Conference Series Material Science and Engineering*, 502:012194.

[2] Barbri H., (2020). *Mechanische und thermische Simulatio eines kryogenen 37-Element Phased-Arrays mit einer tiefsten Temperatur biz zu 4K zur Steigerung der Empfängersensitivität zur Detektion von Weltraumschrott*, Master Work, Hochschule für Technik und Wirschaft Berlin, in German, (*Mechanical and thermal simulation of cryogenic 37-element phased array antenna with lowest temperature till 4K for increasing of receiver sensitivity for detection of space debries*).

CONCLUSION

In the present work, an overview of recently developed radars and communication systems with the superconducting components is given.

State-of-the-art on the development of superconducting components like antennas, filters, limiters, analogue and digital SQUID and DROS is also discussed.

Special attention is given to two promising innovations, namely, electrically small active antennas based on design of bi-SQUID/SQIF blocks, as well as digital electronics working on *Rapid Single Flux Quantum* (RSFQ) logic.

An improvement is possible by application of very narrow 0.5% (or 1% relative) superconducting tunable filter, which is installed in parallel to 200 MHz superconducting filter before LNA. Due to reduction of bandwidth, the system can have 6 dB *signal-to-noise-ratio* improvements.

Application of superconducting antennas with active impedance matching and active antenna tuning could give benefits.

Electrically small active antennas based on bi-SQUID/SQIF blocks can lead to detection of small signals and have relative wide dynamic range. Such antennas are promising, although due to limited market, only very few firms or scientific institutions could produce them.

HYPRES Inc. developed digital receiver systems based on operation of the RSFQ logic. Receivers have very flexible operation, and could be upgrated/updated to more complex systems.

Unfortunately, due to very limited information on the performance of receivers in the opening literature sources, only rough estimations on real operation or on application possibilities could be found and are presented in the present book.

Several other cases, e.g., i) digital SQUID/DROS; analog SQUID, ii) superconducting filters and antenna as well as digital transmitter (direct digital synthesizer); iii) SQUID-based downconverter; iv) and Single-Flux-Quantum computer, have been also shortly mentioned. These developments could be promising for the future, but for the time being, they still need further R&D activities in order to bring significant benefits in comparison to the conventional components or systems.

In summary, the cryogenic front-end receivers bring some advantages over conventional systems at the expense of more complex overall design and increased investments. In which case the complexity and high costs are justified – depends on each application. Without detailed discussion, it is quite difficult to compare all pros and contras of applications of cryogenic front-end receivers.

Appendix A: "Most Significant Change to Satellite Communications Worldwide in 30 Years"

In this appendix, screen-shots from article: "Most Significant Change to Satellite Communications Worldwide in 30 years" http://www.ccas-web.org/superconductivity/wireless/ . See also brochure of CCAS: Coalition for the Commercial Application of Superconductors (2014) under the chapter: *„Applications in Wireless Communications."*

Superconductivity: Applications in Wireless Communications

Superconductor devices, both analog and digital, offer unique advantages for RF (wireless) applications due to ultra-low dissipation and distortion as well as intrinsic (quantum) accuracy. The U.S. Army called their new receiver chip the "most significant change to satellite communications worldwide in 30 years." Superconductor RF filters with superior interference cancellation have been employed in cellular base stations, enabling wider range and fewer dropped calls. Fourth generation All-Digital Receivers (ADRs) for U.S. defense deliver dramatic improvements in performance, cost and efficiency for SIGINT, EW and SATCOM systems. ADRs are deployed in several government programs, addressing the most difficult national RF challenges.

HTS Filters

Cellular base stations receive relatively weak mobile signals with very sensitive antennas and receivers on many channels simultaneously. To perform this task and deal with the tremendous RF interference in the environment some systems employ superconductor RF filters that cancel much of this interference. These filters increase the range of handset due to the superior performance of these ultra-sharp filters. This technology has been commercially deployed in more than 10,000 base stations with over 400 million hours of run time. The cryogenically cooled filters have been exposed to the worse elements and proven very reliable, with Mean-Time-Between-Failure (MTBF) greater than 1 million hours. In an industry where reliability and uptime are the number one goal, these HTS filters have proven that superconductors improve our everyday life.

All-Digital Receivers

While HTS filters have demonstrated unique capabilities in analog (non-digital) RF applications, the true revolution is one that takes advantage of the intrinsic linearity and quantum accuracy of digital superconductor devices (chips) to produce the world's highest performing analog-to-digital converters. Much like digital CDs and digital television provide superior performance, All-Digital Receivers deliver superior performance to all wireless applications. The crux of the improvement is in the unique ability of superconducting analog-to-digital converters to digitize a wide swath of spectrum without the need for analog pre-processing. As a result, significant portions of the conventional RF system are completely eliminated, reducing size and power while significantly improving receiver sensitivity performance. In addition, the manipulation of the digital data enables unparalled flexibility in the signal processing that detects, characterizes and decodes incoming RF signals. These all-digital superconductor receivers have been successfully integrated into legacy wireless systems with the latest cryogenic cooling and systems engineering innovations. From radio astronomy to the latest electronic warfare systems, digital superconductor RF has significantly pushed the state-of-the-art.

"Most Significant Change to Satellite Communications Worldwide in 30 Years"

An early demonstration of the capability if superconducting All-Digital Receivers was carried out by the U.S. Army, where an X-Band satellite downlink was accomplished with a single ADR system that replaced many racks of legacy receiver components. The ability to directly digitize wideband X-Band SATCOM signals (an important military band) proved so impressive it led to the proclamation that the "most significant change to satellite communications worldwide in 30 years" had been achieved.

Today's All-Digital Receivers are packaged in ½ of a 19-inch electronics rack, completely free of liquid cryogen and capable of being equipped to digitize multiple GHz of instantaneous bandwidth virtually anywhere from DC to "W" Band. The modular design can be customized for a variety of applications and is upgradable and programmable with relative ease.

To complement the All-Digital Receiver and address certain applications, a Cryogenic Analog RF Module (CARM) has been recently developed to provide RF amplification with extremely low noise. Initially designed for Ka-Band, the module features a compact cryogenic cooler and can be modified for a wide range of RF bands.

The heart of the All-Digital Receiver system is the superconducting integrated circuit (IC). This chip circuit is made similarly to semiconductors, but the key material is Low Temperature Superconductor (LTS) niobium vs. silicon. The chip - less than half the size of a penny - employs Josephson junctions that make up Rapid Single Flux Quantum (RSFQ) logic circuits that move picosecond-duration magnetic pulses vs. electrons. These LTS ICs need to be cooled to approximately 4 degrees Kelvin compared to 70 degrees Kelvin for analog HTS materials. New innovations in small, off-the-shelf, cryogen-free cryocoolers provide compact cooling platforms for packaging and deploying All-Digital Receivers.

Issues and Recommendations

All-Digital Receivers (and accompanying Transmitters and Transceivers) have been largely developed by and for defense and other government programs with the most difficult RF challenges. Continued government program support is needed to fully exploit the performance capabilities of this technology. Most commercial RF applications are highly commoditized with very large volumes and digital superconductors are still not cost competitive for many applications that otherwise would benefit from significantly improved performance.

Appendix B: Application of 4th Generation ADR with Multiple Channels for Naval Radio Frequency Systems

Department of the Navy SBIR/STTR Transition Program

DISTRIBUTION STATEMENT A. Approved for public release. Distribution is unlimited.

ONR Approval #43-4388-18

Topic # N142-118

Wideband RF Digitizer with Integrated Filter

HYPRES, Inc.

Who

SYSCOM: ONR

Sponsoring Program: SPAWAR PMW 120, PMW 150, PMW/A 170, PMW 770

Transition Target: Shipboard Signals. Exploitation Equipment next generation, Navy Multi-band SATCOM Terminal

TPOC: Dr. Deborah Van Vechten deborah.vanvechten@navy.mil

Other transition opportunities: Multifunctional Information Distribution System (MIDS) Link 16, Magnetic Resonance Imaging (MRI) systems for Traumatic Brain Injury

WHAT

Operational Need and Improvement: Naval radio frequency (RF) systems have a requirement to improve spectrum utilization by implementing high speed broadband analog to digital converters. Although great strides were made with the third generation Modular, Multi-Function Digital-RF Receiver systems in the hands of government labs, attention is now focused on the fourth generation, first delivered in 2018. The ability to simultaneously capture a wide range of signals of interest is now primarily limited by the immaturity of integration of the data output into appropriate backend DSP.

Specifications Required: This phase II is part of the program plan to develop a wideband receiver for applications such as electronic warfare. The objective of this Phase II is to integrate ADCs together with an analog filter/multiplexer as an integrated Nb technology single chip. If more capability is required, the system can be scaled up by combining chips into a multi-chip module (MCM).

Technology Developed: Technology developed utilizing the HYPRES advanced digital RF receiver (ADR) will eliminate many components,enhance toleration of co-located high power transmitters and enable dominance of the RF spectrum for naval multi-function RF applications, such as communications, Electronic Warfare (EW), Intelligence, Surveillance, and Reconnaissance (ISR) and Radar.

Warfighter Value: These Modular, Multi-Function Digital-RF Receiver systems maximize signal reception for communication, ISR and EW systems, enabling full broadband spectrum monitoring and precise emitter identification (ID) in a smaller form factor. Direct digitization at RF eliminates traditional analog RF components such as mixers, LO and many filters. High sensitivity allows many LNA to be eliminated. Digital reception allows many signals having different bandwidths and potentially overlapping frequency content to be collected from a single data stream and a single antenna. All these benefits reduce the size, weight, and power of full spectrum access systems by as much as 50 percent while simultaneously allowing facile real time adaptation.

When

Contract Number: N00014-17-C-2001 Ending on: March 16, 2019

Design and develop a multi-ADC chip and a digital RF receiver concept: 1st QTR FY18

Develop Cryocooled Testbed for Receiver chip: 2nd QTR FY18

Design and develop Nb analog filter: 2nd QTR FY18

Option 1-Develop receiver MCM: 2nd QTR FY20

Option 2-Build Digital-RF Receiver with Filter ADC SoC MCM: 1st QTR FY20

How

Projected Business Model: The output of this Phase II will result in a product that will be incorporatedinto the HYPRES digital RF receiver product family which in turn will be inserted into military systems directly and through other DoD prime contractors. HYPRES has worked with L-3 Communications, Argon ST, and ViaSat on digital-RF receiver projects in the past.

Company Objectives: HYPRES seeks program office support for completion of development, test and evaluation, and transition to Ship's Signals Exploitation Equipment (SSEE) and Navy Multiband Terminal (NMT) as examples of relevant programs of record. HYPRES also seeks relationships with prime contractors, DoD program offices and equity investors for transition to major communications, EW, ISR and radar acquisition programs, and to the commercial market.

Potential Commercial Applications: HYPRES has identified commercial applications in MRI systems (ONR/OSD SBIR phaseI/phase II contracts), avionics equipment and commercial SATCOM.

Contact: Michael DeZego, Director, Business Development mdezego@hypres.com 321-427-5293

Other sources: Rajesh Uppal, January 21, 2020, International Defense, Security & Technology (IDST)

"Radar: Superconductor electronics can dramatically enhance anti-ship missile defense radars. Emerging threats include sea-skimming missiles that reflect very small fraction of the radar signal. The challenges to radar receivers are that they must distinguish these small echoes from the huge background clutter of waves, rain, jammers and mountains on the shore in real time. Superconductivity enable the highest dynamic range digitizers and the smallest digits are essential to detect the most elusive threats. High dynamic range superconductor-based electronics provide the most advanced technology and simplify the receiver, thereby making these life-saving sensors and detectors affordable for a wide variety of Navy ships. CCAS recommends that programs in support of such major technology advances and military advantages should be expanded as in combat situations settling for second best is not good enough."

(see also CCAS brochure "Superconductivity: Present and Future Applications)

"Passive Microwave and Millimeter-wave Components enabling increased detection range and discrimination in clutter. This could lead to development of MMW radar would embody HTS materials in its filters, transmission lines, phase shifters and possibly reflectors"

"Staring Infrared Focal Plane Array sensors incorporating superconducting electronics permitting significant range and sensitivity increases over current scanning IT sensors. This could lead to improved Space Surveillance System through development of high resolution and low power consumption IR focal plane array by combing detectors using existing extrinsic silicon materials with signal processors employing LTS materials"

About the Author

Sergiy Putselyk earned his BS (1999) and MS (2001) in Radiophysics from Taras Shevchenko National University of Kyiv, Ukraine, as well as PhD (2005) in Applied Physics from University of Bayreuth, Germany, and ME (2006) in Process Engineering from Offenburg University of Applied Sciences, Germany. Dr. Putselyk's research interests include design and construction of medium to large scale set-ups or facilities for scientific and industrial needs as well as project management. He gained practical experience in several fields, e.g., cryogenics, vacuum and high frequency techniques, super- and normal conducting magnets, infrastructure construction for scientific/semi-industrial purposes, measurements and industrial automation systems.

INDEX

J

L

M

N

O

P

Q

R

S

T